Van Blerkom/Motta
The Cellular Basis of Mammalian Reproduction

The Cellular Basis of Mammalian Reproduction

Jonathan Van Blerkom, Ph.D.
Department of Molecular, Cellular
and Developmental Biology
University of Colorado
Boulder, Colorado USA

Pietro Motta, Ph.D.
Department of Anatomy
Faculty of Medicine
University of Rome
Rome, Italy

235 Electron Micrographs on 85 Plates

Urban & Schwarzenberg · Baltimore-Munich 1979

Urban & Schwarzenberg, Inc.
7 E. Redwood Street
Baltimore, Maryland   21202
U.S.A.

Urban & Schwarzenberg
Pettenkoferstrasse 18
D-8000 München 2
GERMANY

Library of Congress Cataloging in Publication Data

Van Blerkom, Jonathan, 1947–
   The cellular basis of mammalian reproduction.

   1. Mammals--Reproduction. 2. Mammals--Cytology. 3. Generative organs, Female.
I. Motta, Pietro, joint author. II. Title. [DNLM: 1. Reproduction. 2. Mammals.
3. Microscopy, electron. 4. Cytology. WQ205 V218c]
QL739.2.V36   599'.01'6   78–10230

ISBN 0-8067-2041-7 (Baltimore)
ISBN 3-541-72041-7 (München)
Printed in Germany by Kastner & Callwey, München

This book is dedicated
to the memory of
Professor Giuseppe Motta
(1902–1966)
Chairman of the Department of Obstetrics & Gynecology
University of Messina
whose scientific works and conversations
are to a large extent
embodied in this book

To our parents and wives

Contents

Section I
The Ovary, Ovulation, and the Female Reproductive Tract

Section II
The Gametes and the Preimplantation Stages of Pregnancy

Preface

Since bisexual reproduction in mammals is so central to the continuation of the species, one might assume that investigations on the cells and phenomena involved would by now have answered most of the important questions. And yet quite the opposite appears to be the case, as dramatically demonstrated by the contents of this unique book. That there is ever more to learn and understand is especially emphasized in this instance by the application of new instruments and techniques.

Here the authors have made generous and effective use of the unusual capabilities of the scanning electron microscope to study the minute details of cell and tissue surfaces and the interrelations of these surfaces in the processes of reproduction. The great part of this information has not before been presented in book form and some of it has not previously been published anywhere. Much of the latter is drawn from several years of study on the part of the authors themselves.

The reader will recognize in the text a valuable blending of physiology, morphology, and molecular biology which, together with abundant literature references, makes the book useful for both the beginning student and the established investigator of reproduction. A field as currently active as reproductive biology needs occasional books of this nature to keep creative thought and research focussed on the important problems and their solution.

If kept up-to-date with successive editions, this book and its authors will perform a long-lasting service to the development of this important field of biology.

KEITH R. PORTER

Acknowledgements

We are indebted to Prof. Meredith N. Runner (Boulder, Colorado), Dr. Robert W. McGaughey (Tempe, Arizona), Dr. Martin H. Johnson (Cambridge, England) and Mr. Hobart Bell for reading the manuscript of the book and for providing many useful comments and suggestions. We are also grateful to Ms. Pamela Akinson and Dr. S. Corner for help in the preparation of the manuscript, and to all of those at Urban & Schwarzenberg for the opportunity to produce the book. Special thanks are accorded to the investigators who generously provided photographs from their published works, and to the publishers of the following journals and books for permission to reprint published material.

Plate 2 B:	P. Motta: Acta anat. (Basel) *90:* 36–64, 1974
Plate 11 B:	P. Motta, Z. Takeva and D. Palermo: Acta anat. (Basel) *78:* 591–603, 1971
Plates 15 B, 16 A:	P. Motta and J. Van Blerkom: J. Submicr. Cytol. *6:* 297–310, 1974
Plate 19 C:	P. Motta and L. J. A. DiDio: J. Submicr. Cytol. *6:* 15–27, 1974
Plates 20 A, 21, 22, 24 A, B, D, 25 A, 26:	P. Motta and J. Van Blerkom: Am. J. Anat. *143:* 241–264, 1975
Plates 25 B, 28, 29 A, 31 C, 33, 37:	J. Van Blerkom and P. Motta: Cell Tiss. Res. *189:* 131–153, 1978
Plates 49, 50, 55:	P. Motta and J. Van Blerkom: Cell Tiss. Res. *163:* 29–44, 1975
Plates 56 A, 57 A, 58 B:	D. D. Eager, M. H. Johnson and K. W. Thurley: J. Cell Sci *22:* 345–353, 1976
Plate 60:	J. Van Blerkom and M. N. Runner: J. Exp. Zool. *196:* 113–123, 1976
Plates 63 B, 68 A, B, E, 69, 70, 79 A, C:	J. Van Blerkom, C. Manes and J. C. Daniel: Devel. Biol. *35:* 262–282, 1973
Plates 72 C, 82 B, 83 A:	P. G. Calarco: In: Scanning Electron Microscopic Atlas of Mammalian Reproduction, (E. S. E. Hafez, ed.) Springer-Verlag, Berlin, 1975
Plate 74:	From Drs. T. Ducibella, E. Anderson, T. Ukena and M. J. Karnovsky and published by T. Ducibella: In: Development in Mammals, vol. 1, (M. H. Johnson, ed.). North Holland, Amsterdam, 1977.

Finally, much of the work presented in this book has been supported by grants from the National Institutes of Health, United States Public Health Service, the Fulbright Foundation, and from the Consiglio Nationale delle Ricerche.

List of Plates

Introduction

The examination of biological material by scanning electron microscopy has provided new approaches to the study of cell structure and function and a wealth of information relevant to the cellular basis of developmental events. A combination of scanning and conventional light and transmission electron microscopy offers the potential of deriving new and fundamental insights into morphological correlates of physiological processes. This combination has proven to be of particular value when applied to the study of mammalian reproduction. It is the intent of this book to describe the results of such an approach and to present a comprehensive account of selected aspects of mammalian reproduction which have heretofore not been reviewed in this manner. Therefore, the primary emphasis has been placed on the morphodynamic processes related to (1) the ovary, (2) ovulation, (3) the female reproductive tract, (4) the attainment of the fertilizable state of the gametes, and (5) the preimplantation stages of pregnancy. The interpretations of many of the observations are original and, in some cases, speculative. Nevertheless, it is our opinion that many of the recent experimental findings and concepts that are discussed and that represent the work of numerous investigators indicate exciting, critical, and controversial areas in the study of reproduction. We have attempted to provide a detailed bibliography of original and contemporary references which have been published in a wide variety of journals and books. Because we are well aware of the many reports in the literature that could have been included but, due to limitations in space, have not, we apologize to those investigators whose published works have not been cited. The information contained in this book should be of use to students and researchers concerned with reproduction, cellular and developmental biology, anatomy and histology, and to those involved in clinical and veterinary medicine.

Section I

The Ovary, Ovulation, and the Female Reproductive Tract

The purpose of this section is to provide a detailed survey of the fine structural aspects of the female reproductive system during the various phases of the estrous cycle. Much of the current information regarding cellular aspects of the reproductive system, as observed by transmission and scanning electron microscopy (TEM and SEM), have come from laboratory animals (rats, mice, rabbits, hamsters, etc.). In addition, a good portion of this information is derived from animals maintained on hormonal regimens or induced to ovulate following direct, exogenous hormonal intervention. Only a limited number of studies dealing with human material is available (see *Ferenczy* and *Richart,* 1974), and frequently, the findings of some of these studies have questionable relevance to the "normal" physiological condition because specimens were obtained from women treated for various diseases of the reproductive tract.

However, many of the morphophysiological changes of the human reproductive tract during its normal reproductive cycle have been shown to be similar to changes that occur in the genital tracts of other vertebrates, especially common laboratory animals. Although the information contained within the following three chapters was derived from laboratory animals undergoing *natural* reproductive cycles, many of the basic processes may be assumed to be quite similar to the human situation.

The reproductive system of the mammalian female consists of a pair of ovaries, related ducts (oviducts, uterus, vagina), and external genitalia. Descriptions of the mammary glands and external genitalia have been omitted in deference to a more detailed discussion of the morphophysiological changes occurring within and on the surface of the cells of the ovaries and related ducts during the estrous cycle.

1 The Ovary Prior to Ovulation

1.1. General Morphology

The mammalian ovaries are paired and highly complex organs capable of the following major functions: (1) the production of female gametes (oocytes) during the entire reproductive life span of the organism, (2) the secretion of specific steroid hormones (e.g., estrogens and progesterone) essential for all sexual events occurring within the genital tract (estrous cycle, menstrual cycle, pregnancy), and (3) the hormonal regulation of the postnatal growth of the reproductive organs and the development of the secondary sexual characteristics. Except for functional or pathological changes, the ovaries are approximately equal in size and are attached by the mesovarium to the broad ligament.

If cut in midsection, the ovary, even to the unaided eye, consists of an outer zone, the cortex, which is grossly separated from an internal zone, the medulla. Both the cortex and the medulla are composed of a stroma of connective tissue (stroma ovarii) containing collagen fibers, fibroblasts, scattered smooth muscle cells, and some elastic fibers. Usually, there is no clear line of demarcation between cortex and medulla, and some cortical elements extend into the medulla and vice versa (Mossman and Duke, 1973). Although the terms cortex and medulla are used to denote rough topographical levels within the ovary, the delineation of these regions results from the differential organization of the stroma. The stroma of the cortex is a dense connective tissue rich in cells, whereas the stroma of the medulla is composed of loose connective tissue in which are embedded large nerves, lymphatics, and numerous blood vessels. These nerves and vessels take a rather tortuous path through the organ, and both enter and exit the ovary via a sinus, the hylus.

Concentrated in the cortex of the ovary are numerous small, primary follicles (containing oocytes in a "quiescent" state) as well as a reduced number of comparatively larger follicles. During each estrous or menstrual cycle, a variable number of follicles enter into a phase of rapid growth and development culminating in the process of ovulation (Chapter 2). Near the periphery of the ovary, just beneath the superficial epithelium, the stroma forms a dense, fibrous, connective tissue layer which covers the entire organ, the tunica albuginea. For a comprehensive treatment of the fetal development of the ovary, or for a detailed survey of comparative aspects of ovarian morphology, the reader's attention is directed to the works of Watzka (1957), Mossman and Duke (1973), Balboni (1976) and Zuckerman and Weir (1977).

Prior to a discussion of the fine morphology of the ovary, it is important to note that any study of a chemically fixed ovary, whether by light microscopy or by transmission (TEM) and scanning electron microscopy (SEM), has to be considered "nothing more than a still frame out of a cinema of cyclic anatomical and physiological events of what seems to be the most rapidly and cyclically changeable organ in the body" (Greep, 1963). This observation should be kept in mind not only in reference to the luteofollicular complex (Chapter 2), but also when attempting to comprehend changes in other ovarian structures such as the thecal and interstitial glands and, possibly, the stroma and superficial epithelium as well.

1.2. The (Germinal) Superficial Epithelium

The exposed surface of the mammalian ovary is covered by a continuous layer of mesothelial cells commonly referred to as the "germinal" superficial epithelium, as it was once mistakenly believed that these cells were capable of generating new germ cells (oogonia) (Plate 1). The superficial epithelium is usually composed of a single layer of polyhedral (columnar or flattened) cells which are tenuously attached to the tunica albuginea (Plates 2 A, 7 C, 7 D). Generally, the free surfaces of these cells are covered with a large number of microvilli, which can attain an average density of between 300 and 350 units per square millimeter of cell surface (Plates 1, 4). With some minor variations in length and diameter, these microvilli are rather uniform in appearance (Plates 1 B, 1 D). A single cilium, typically located in the central portion of the cell, is often observed projecting above surrounding microvilli (Plates 1 E, 7 D). At high magnification, the exposed surfaces of the microvilli have a rather rough appearance owing to the presence of an amorphous material which is similar in texture to the glycocalyx coat present on microvilli in other tissues (the intestinal epithelium, for example). Areas of superficial cells devoid of microvilli display a number of small pits and cavities which open into cortical caveolae (Plate 1 E).

A characteristic feature of the ovarian epithelium is the high degree of variability in the density of microvilli on superficial cells (1) located in different regions of the same ovary, (2) of ovaries of different mammals, (3) of different ages, and (4) of ovaries in different phases of the estrous cycle. Furthermore, by SEM, cells of the superficial epithelium display several features likely related to the phase of the cell cycle; these features include branching evaginations, blebs, ruffles, or lamellipodia, as well as cells with relatively smooth surfaces (Porter, Prescott, and Frye, 1973). Cells in which either blebs and/or lamellipodia are evident have a reduced population of microvilli, thus suggesting that microvilli may not be a permanent aspect of these cells but rather may be continually replaced by other surface features. Changes in morphology and architecture that seem to be associated with the cell cycle are undoubtedly surface manifestations both of the phase of the reproductive cycle and of events occurring within the ovary. The various morphological and organizational aspects of the ovarian epithelium, which are likely related to the cyclic and dynamic nature of this organ, make this mesothelium somewhat different from other mesothelial linings, including the peritoneal and visceral mesothelia (Andrews and Porter, 1973; Mossman and Duke, 1973).

The occurrence of a single cilium on some cells of the superficial epithelium is a feature common to mesothelial surfaces (Motta, Andrews, and Porter, 1977). Cilia, which are observed with difficulty by TEM, are encountered not only on the epithelium of the ovary (Plates 1 E, 7 D) (Sato, 1965; Wischnitzer, 1965; Tanaka, Sugawara, and Hafez, 1975; Anderson, Lee, Letourneau, Albertini, and Meller, 1976) but also on cells within the ovary itself (Motta, Takeva, and Palermo, 1971) and primarily on follicle cells (Adams and Hertig, 1964; Motta, 1965a; Stegner, 1967). It seems most probable that ovarian cilia are nonfunctional, rudimentary organelles. It is nevertheless an intriguing possibility that they may have some as yet undefined role in the coordination of physiological processes during follicular growth or, as previously postulated, function in a chemoreceptive capacity (Motta, 1965a).

On a more macroscopic level, the surface of the ovary is frequently evaginated into a series of villous-like projections or papillae which, depending on the species, may vary widely in number, size, and distribution (Dabelow, 1939; Harrison and Matthews, 1951). As an example, rabbit ovarian papillae are generally quite numerous, occasionally visible to the unaided eye, and frequently assume intriguing and unusual arrangements (Plates 3, 6 A, 7 A, 7 B) (Cherney, Motta, and DiDio, 1973; Motta and Van Blerkom, 1975). In other mammals, including humans (Dabelow, 1939; Sternberg, 1963), papillae are comparatively smaller and reduced in number at least under normal physiological conditions (Plate 4) (Jensen and Norris, 1972).

In several species, areas of the superficial epithelium are often invaginated into subjacent, cortical layers. These invaginations form small, well-defined, simple or ramified crypts and cords (Plates 2, 5, 6 B) (Harrison and Matthews, 1951; Mossman and Duke, 1973). As observed by both TEM and SEM (Plates 2, 5, 6 B), the fine morphology of the cords and crypts in the mature ovary of the rabbit clearly demonstrates that these structures are true infoldings of the superficial epithelium. Crypts are hollow, tubular invaginations in which the lumen opens directly to the surface of the ovary (Plates 2 A, 5, 6 B). The cells on the surface of the ovary that surround the openings of the crypts are similar in appearance to other cells of the superficial epithelium; some of these cells possess numerous microvilli, whereas others are relatively flattened or elongated, display few microvilli, and may have either a smooth surface or a surface containing blebs (Plates 5, 6 B) (Wischnitzer, 1965; Weakley, 1969; Motta, Cherney, and DiDio, 1971; Papadaki and Belby, 1971; Ferenczy and Richart, 1974; Jeppesen, 1975; Anderson, Lee, Letourneau, Albertini, and Meller, 1976; Ludwig and Metzger, 1976). Crypt cells are polyhedral elements (columnar or cuboidal) with a lobulated nucleus and a cytoplasm characterized by numerous free ribosomes, polysomes, a Golgi complex, relatively small mitochondria, and a few narrow cisternae of rough-surfaced endoplasmic reticulum (Motta, 1974 a). The exposed surfaces of crypt cells that face the lumen display numerous microvilli, an occasional cilium, blebs, and ruffles (Plates 2 A, 5, 6 B). Crypt cells are interconnected both by interdigitations of the plasma membrane and by junctional complexes.

Some invaginations of the superficial epithelium terminate as small, irregular, fragmented cords which, at higher magnification, appear as "nests" of epithelial-like cells. These nests lie in close proximity to developing follicles (Plate 2 B) and/or to isolated clusters of interstitial gland cells (discussed below). Generally, owing to a more solid configuration, cord-like structures do not have a lumen, and the cells composing the cords do not possess either large cellular projections or a sizeable population of microvilli. At the fine structural level, cord cells contain lipid droplets, numerous free ribosomes, lysosomes, a Golgi complex, and a reduced number of elements of smooth-surfaced endoplasmic reticulum. Intercellular contact is maintained through tight and/or gap junctions (Plate 2 B).

Ultrastructurally, cord cells appear quite similar to granulosa cells associated with developing follicles in regions adjacent to the cords (*Unsicker,* 1971) and also to some groups of differentiating interstitial cells (*Mori* and *Matsumoto,* 1970; *Motta,* 1974 a). Furthermore, cord and crypt cells are morphologically similar to the cells of the superficial epithelium. On strictly fine structural criteria, one interpretation of these observations suggests that the superficial epithelium, with its known proliferative abilities, may serve as an important and continuous source of these and other ovarian components present during adult life. In this regard, the superficial epithelium of the ovary may represent a true "germinative" epithelium (*Motta,* 1974 a). Clearly, other criteria are required to support this hypothesis.

The presence of both invaginations and evaginations on the ovarian surface demonstrates the dynamic nature of this superficial epithelium. It has been shown that the incidence, number, size, and degree of development of the superficial structures (crypts, cords, and papillae) may be influenced by the levels of circulating pituitary gonadotropins and/or ovarian steroid hormones (*Harrison,* 1962; *Jensen* and *Norris,* 1972). Therefore, it seems likely that these structures are related to the reproductive cycle of each species in which they occur. Consequently, the formation of papillae, cords, and crypts may be simply a proliferative expression of the superficial epithelium under normal, cyclic conditions. The demonstration of submicroscopic changes on the surfaces of the cells of the ovarian epithelium (microvilli, blebs, ruffles), as well as the more macroscopic elaboration of papillae and crypts, strongly indicates that the superficial epithelium undergoes a cyclic process similar to that which occurs in other parts of the ovary and in other regions of the genital tract during the reproductive cycle (*Motta,* 1974 a). Finally, as further evidence of the dynamic nature of the superficial epithelium, it has recently been demonstrated that the superficial cells, especially in the basal areas of the ovary, contain numerous coated and noncoated endocytotic caveolae (*Anderson, Lee, Letourneau, Albertini,* and *Meller,* 1976). These caveolae, which are capable of internalizing exogenously administered horseradish peroxidase, are thought to be an efficient transport mechanism for the removal of material from the peritoneum and interstitium of the ovary.

1.3. Follicular Development

During the course of a single estrous or menstrual cycle, a variable number of follicles undergo a process of rapid growth and development culminating in the liberation of an oocyte(s) at ovulation (Chapter 2). In the human, typically only one oocyte is discharged per cycle, whereas in other mammals, multiple ovulations are common. In the sexually mature animal, ovarian follicles in various stages of growth are embedded in the cortex of the ovary.

At the outset of growth, follicles are approximately 40 to 50 μm in diameter and contain an oocyte (about 15 to 30 μm in diameter) surrounded by a single layer of irregularly shaped, flattened cells (Plate 8). These follicles are termed primordial or primary, unilaminar follicles, and the oocyte, a primary or quiescent oocyte. Small follicles are usually separated from the surrounding stroma by a thin basal lamina which increases in thickness considerably during subsequent stages of follicular growth (Plates 8, 9). With continued growth, additional layers of cells surround the oocyte, and such follicles are termed secondary or growing follicles. No appreciable fluid is present in the intercellular spaces of secondary follicles. The accumulation of liquor folliculi and the resultant appearance of an antrum occur during the latter phases of growth, and follicles that have attained such a state are commonly known as tertiary, vesicular, Graafian, and preovulatory follicles. Inherent in the use of the traditional descriptive terms is that follicles which differ slightly in development may be classified together. A more precise system of classifying oocytes and follicles has been developed by *Pedersen* and *Peters* (1968) and is based on (1) the size of the oocyte in follicles at different stages of development, (2) the size of the follicle as defined by the number of cells constituting the follicular envelope, and (3) the morphology of the follicle (see also *Pedersen,* 1969; and *Peters,* 1969).

The initiation of the growth of the oocyte is most likely regulated within the ovary, such that the number of oocytes that enter the growth phase is related to the size of the pool of nongrowing oocytes (*Krarup, Pedersen,* and *Faber,* 1969; *Peters, Byskov, Lintern-Moore, Faber,* and *Andersen,* 1973). Typically, both oocyte and follicle grow coordinately and, in the process of development, progress through a sequence of definable morphological stages (*Pedersen* and *Peters,* 1968). In the mouse, oocyte growth is completed before the formation of the antrum, and, thus, the major portion of follicular growth occurs subsequent to the completion of growth by the oocyte (*Brambell,* 1928). In any event, the volume of both follicle and oocyte increases markedly, such that at ovulation and depending upon the species, an oocyte may range between 100 and 150 μm in diameter, whereas the follicle may attain a diameter of nearly 12 mm!

For purposes of illustration, the following discussion presents the developmental events associated with the growth of a primary oocyte and the development of a primary follicle, which could lead to the discharge of the mature oocyte at ovulation (Chapter 2) or to the atretic modification of both oocyte and follicle (see section 1.3.4).

1.3.1. The Maturative Growth of the Oocyte

The oocyte of a unilaminar, primary or type 3 follicle (*Pedersen* and *Peters,* 1968) is generally considered to be in a "quiescent" state since no major morphological changes take place within these oocytes during the protracted period (in some mammals) from birth to sexual maturity (*Zamboni,* 1972). Meiosis in quiescent oocytes is arrested at the diplotene stage of prophase. The nucleus of a quiescent oocyte is relatively large and vesicular; it contains numerous pores, finely dispersed chromatin, and, usually, one or more large nucleoli. The nucleoli are reticulated, and both fibrillar and granular elements are evident. Most of the organelles in these oocytes, such as the Golgi complex, lysosomes, mitochondria, and the endoplasmic reticulum are concentrated in a juxtanuclear position (Plate 8). This region, which is Feulgen positive, is evident at the level of the light microscope, is usually crescent-shaped, and is known as the Balbiani vitelline body (*Raven,* 1961; *Weakley,* 1967; *Hertig,* 1968). The preovulatory growth and meiotic maturation of a stimulated "quiescent" oocyte is accompanied by dramatic changes at the subcellular level (*Yamada, Muta, Motomura,* and *Koga,* 1957; *Sotelo* and *Porter,* 1959; *Odor,* 1960; *Warten-*berg and *Stegner,* 1960; *Blanchette,* 1961; *Adams* and *Hertig,* 1964; *Weakley,* 1966; *Zamboni* and *Mastroianni,* 1966; *Baca* and *Zamboni,* 1967; *Baker* and *Franchi,* 1967; *Hertig* and *Adams,* 1967; *Pavelka, Friedrich,* and *Caucig,* 1972; and reviewed by *Zamboni,* 1972; *Mestwerdt, Muller,* and *Brandau,* 1977). For example, in the quiescent oocyte, the Golgi complex is a large and extended assemblage of closely packed tubules and vesicles. During the early stages of maturative growth, Golgi elements migrate from their juxtanuclear position to a subcortical location and, in the process, subdivide into numerous smaller components (*Adams* and *Hertig,* 1964; *Zamboni* and *Mastroianni,* 1966; *Szollosi,* 1967). Since the migration of the Golgi components occurs simultaneously with the deposition of the zona pellucida and usually precedes the appearance of dense cortical granules, it has been suggested that the Golgi complex may have a central role not only in the formation of the zona pellucida (see section 1.3.3) but also in the formation of cortical granules (*Zamboni* and *Mastroianni,* 1966; *Szollosi,* 1967). The development of cortical granules usually begins during the latter phases of preovulatory growth and, in some species, persists in the ovulated oocyte prior to fertilization (*Zamboni,* 1970). The contents of the cortical granules are discharged from the cytoplasm at the time of sperm penetration, and their biochemical and histological properties support the hypothesis that they are a type of oocyte lysosome (*Dalcq,* 1963; *Zamboni,* 1970) (see Chapter 5).

Rather intriguing structures present in both quiescent and maturing oocytes resemble the fine structural appearance of nucleoli. These structures, usually referred to as nucleolar-like bodies or as nuage, are observed throughout the cytoplasm as discrete entities and/or as "cementing" material situated in the interstices of mitochondrial clusters (Plate 8 and Chapter 5) (*Adams* and *Hertig,* 1964; *Hope,* 1965; *Odor,* 1965; *Weakley,* 1969; *Motta* and *Van Blerkom,* 1974a). Nuage material has been described not only for oocytes, but also for preimplantation embryos and male and female germ cells (*Motta* and *Van Blerkom,* 1974a). The fine structural appearance of nuage is similar to (1) inclusions in the spermatocytes of numerous species, where they are termed chromatoid bodies, (2) the polar granules in the primordial germ cells and early embryos of *Drosophila melanogaster* (*Mahowald,* 1962), and (3) the germinal granules in the germ of *Rana pipiens* and *Xenopus laevis* (see *Smith* and *Williams,* 1975 for a review). *Eddy* (1974) has suggested that this material may represent the mammalian germ cell determinant. However, more detailed cellular and biochemical studies are required to support this hypothesis.

During the maturative growth of the oocyte, numerous microvilli are expressed in the surface of the oocyte, and many intermingle freely with the elongated processes of the surrounding follicle cells. These processes branch in an irregular fashion toward the oocyte surface and have been observed to penetrate the cortex of the oocyte (*Zamboni,* 1974). The more common association of oocyte and follicle cell process is by tight and gap junctions, as well as by junctional complexes (see *Zamboni,* 1974 for a comprehensive review, and also the following section). Other organelles, such as rough-surfaced endoplasmic reticulum (RER), mitochondria, and ribosomes, increase significantly in number and become uniformly distributed throughout the cytoplasm (i.e., from their polarized, juxtanuclear position characteristic of the quiescent oocyte). In the mouse, rat, and hamster oocyte, fibrous material, forming extensive lattice-like structures or plates of parallel sheaths, appears rather suddenly in the cytoplasm of growing oocytes (see *Zamboni,* 1972, and Chapter 5). During the preovulatory growth of a mouse oocyte, the total volume of the oocyte increases approximately 40-fold, while the volume of the nucleus and nucleolus increases approximately 6- and 90-fold, respectively (*Schultz* and *Wassarman,* 1977a).

At the molecular level, numerous studies have demonstrated that the growing oocyte is actively synthesizing both protein and RNA (*Oakberg,* 1968; *Baker, Beaumont,* and *Franchi,* 1969; *Moore, Lintern-Moore, Peters,* and *Faber,* 1974; *Rodman* and *Bachvarova,* 1976, *Wassarman* and *Letourneau,* 1976; *Schultz* and *Wassarman,* 1977a). The period of rapid growth and fine structural reorganization "ends" with an oocyte, which is competent to resume meiosis (meiosis is arrested at the diplotene stage of prophase throughout growth), located in a tertiary or Graafian follicle (type 7 or 8; *Pedersen* and *Peters,* 1968). The period of time during which meiosis progresses from the dictyate (germinal vesicle) stage to metaphase II is known as meiotic maturation, and this terminal phase of oocyte development takes place late in oogenesis – approximately 12 hours prior to ovulation. Accompanying the final meiotic maturation of a fully grown oocyte are qualitative and quantitative changes in the pattern of protein synthesis (*McGaughey* and *Van Blerkom,* 1977; *Schultz* and *Wassarman,* 1977b; *Van Blerkom,* 1977; *Warnes, Moor,* and *Johnson,* 1977; *Van Blerkom* and

McGaughey, 1978). At the cellular level, the oocyte nucleolus condenses into a dense, spheroidal structure containing only fibrillar material. Prior to condensation, oocyte nucleoli are highly reticulated structures in which granular and fibrillar materials are distributed within anastomosing networks of nucleolonemas (*Zamboni* 1972). Additional cellular characteristics of the meiotic maturation of mammalian oocytes are presented in Chapter 2.

1.3.2. Follicle Cell Development

Concomitant with the growth of the oocyte, the flattened granulosa cells of the primordial or unilaminar follicle begin to proliferate rapidly and in the process assume an irregular, polyhedral appearance. The end result of this proliferation is the formation of a multilayered, stratified, epithelial-like structure which surrounds the oocyte – the stratum granulosum or, simply, the granulosa (Plates 9, 11 A). In this stage of growth, follicles are classified as secondary or multilayered follicles (types 4 and 5; *Pedersen* and *Peters,* 1968). Toward the end of the proliferative phase, granulosa cells of multilayered follicles, under the influence of gonadotropins from the anterior pituitary, begin to secrete a fluid, the liquor folliculi, which accumulates in intercellular spaces (Plate 9B) (type 6 follicle; *Pedersen* and *Peters,* 1968). The continued secretion and accumulation of liquor folliculi result in the dissociation of granulosa cells, which subsequently leads to the formation of a large, fluid-filled cavity – the antrum (Plate 9B). When follicles possess a large antrum, they are referred to as vesicular or tertiary follicles or, if quite large, as Graafian follicles (after de Graaf, 1641–73, who first described them) (type 7 or 8; *Pedersen* and *Peters,* 1968). In the Graafian follicle, the continuous accumulation of fluid has both separated granulosa cells closest to the oocyte from other cells in the follicle and, also, has displaced the oocyte with its surrounding mass of follicle cells (cumulus oophorus) to an eccentric position within the antrum (Plate 9B and Chapter 2). The fine structural changes that occur during the growth phase of follicular development are typical of secretory cells with elevated rates of protein synthesis (*Bjorkman,* 1962; *Franceschini, Santoro,* and *Motta,* 1965; *Van Blerkom* and *Manes,* 1974; *Mestwerdt, Muller,* and *Brandau,* 1977). The cytoplasm of the follicle cells contains a dense population of ribosomes and polysomes, Golgi complexes, and numerous cisternae of the RER (Plates 11B, 14C).

Coincident with the initial elaboration of follicular fluid (liquor folliculi) and the rapid proliferation of granulosa cells in the stratum granulosum is the formation of Call-Exner bodies. First identified by *Call* and *Exner* (1875), these structures were originally and erroneously thought to be abnormal ova. More recently, these bodies have been mistakenly considered to be PAS (periodic acid-Schiff) positive material *within* the cytoplasm of granulosa cells (*Hadek,* 1963, 1969). On the contrary, Call-Exner bodies are simply large (10 to 40 microns in diameter), spheroidal, *intercellular* cavities containing material that is histochemically similar to liquor folliculi present in adjacent intercellular spaces (Plate 10) (*Jacoby,* 1962). Call-Exner bodies are limited by a basal lamina-like structure to which granulosa cells are attached in a manner characteristic of epithelial cells (*Motta,* 1965b; *Zamboni,* 1971; *Motta* and *Van Blerkom,* 1974b). The lateral, adjacent surfaces of these cells are attached by tight junctions (*Zamboni,* 1971). These relatively large structures are evident in regions of the granulosa where follicle cells are compacted and follicular fluid is more abundant.

Comparative studies of follicles at various stages of growth and in different species demonstrate two stages in the progressive formation of the Call-Exner bodies. In the first stage, the secretory poles of the granulosa cells are directed toward a common point, which is typically a small intercellular space. During the second phase, the small space(s) becomes filled with fluid, and the original secretory poles become basal poles of granulosa cells behind which a basal lamina-like structure is formed. The appearance of the basal lamina produces a closed cyst, which is typical of the Call-Exner bodies. Occasionally, the contents of Call-Exner bodies are liberated into the antrum of the follicle through a gap in the wall of the cyst. The contents of the cyst appear to blend into the filamentous material that forms the zona pellucida; it has been suggested that the Call-Exner bodies may contain precursors of both the liquor folliculi and the zona pellucida and therefore may represent reservoirs in which this material is stored prior to release into the antrum of a growing follicle (*Zamboni,* 1971). Call-Exner bodies are also observed in atretic follicles (see section 1.3.4) but disappear as the granulosa continues to degenerate.

Numerous studies confirm that the radial arrangement of granulosa cells about the cavity of the Call-Exner body is a general phenomenon of Graafian follicles in the mammalian ovary (*Motta* and *Nesci,* 1969). This arrangement may reflect the "natural propensity" of granulosa cells to become radially arranged as occurs (1) around the oocyte, where they form the corona radiata, (2) during culture *in vitro* (*Harrison,* 1962; *Nicosia* and *Mikhai,* 1975), or (3) in granulosa cell tumors where they are usually referred to as "rosette figures" (*Jensen* and *Norris,* 1972). In the latter two examples, during the rapid proliferation of granulosa cells, these elements become sufficiently detached from one another to reaggregate into hollow spheres in a manner characteristic of the tissue of origin – a spheroidal follicular cavity or the corona radiata (*Motta* and *Nesci,* 1969; *Motta* and *Van Blerkom,* 1978).

Most of the stratum granulosum of a developing follicle contains irregularly polyhedral cells that are coated with an amorphous, viscous material (Plate 11 A). By contrast, the cells that face the antrum are elongated and flattened and are covered with a considerable amount of a filamentous/reticular material. This material is most likely composed of precipitated liquor folliculi and cellular debris, which together form a thick, regular lamina surrounding the antrum. A lamina of this type, localized on the interior surface of the granulosa layer that faces the antrum of a Graafian follicle, was originally described by *Bouin* and *Ancel* (1908) and was termed the membrana limitans interna.

Other follicle cells, especially those located around the oocyte (Plates 12, 13, 14 C) and Call-Exner bodies (Plate 10), have a stellar shape. This somewhat unusual appearance is due to the elaboration of numerous ameboid evaginations of these granulosa cells (Plate 15 A). Microfilaments, 40 to 70 Å in thickness, as well as the contractile proteins actin and myosin, have been identified in these cells and especially in the cellular processes (Plate 15 B and Chapter 2) (*Ansalone, Familiari, Fedele,* and *Motta,* 1973; *Schuchner* and *Stockert,* 1973; *Van Blerkom* and *Manes,* 1974; *Cavallotti, DiDio, Familiari, Fumagalli,* and *Motta,* 1975, *Van Blerkom* and *Brockway,* 1975; *McGaughey* and *Van Blerkom,* 1977). This extensive filamentous network is thought to be responsible for the *in vivo* contractions and pulsations of granulosa cells and also may be involved in the retraction of granulosa cell processes from the oocyte surface at ovulation (see section 1.3.3 and Chapter 2) (*Motta* and *DiDio,* 1974; *Motta* and *Van Blerkom,* 1974b).

The nature of intercellular contact and communication between granulosa cells has been examined by conventional TEM (Plate 11 C) (*Franchi,* 1960; *Bjorkman,* 1962; *Franceschini, Santoro,* and *Motta,* 1965; *Byskov,* 1969; *Zamboni,* 1970; *Espey* and

Stutts, 1972; *Merk, Albright* and *Botticelli*, 1973) and freeze-fracture methods (*Anderson*, 1971; *Albertini* and *Anderson*, 1974). Collectively, these studies demonstrate that the primary means of communication between granulosa cells is by randomly distributed gap junctions. These membrane specializations are of special interest in granulosa cell communication since they increase in both size and number during the growth of the follicle (*Albertini* and *Anderson*, 1974). This observation supports the suggestion of *Motta, Takeva,* and *Nesci* (1971) that gap junctions, as well as other types of intercellular contact, may have, other than mechanical significance, a role in the possible coordination of cellular activities during follicle growth and oocyte development. Also, gap junctions increase in size and number primarily during the phase of follicle growth that is dependent upon pituitary gonadotropin stimulation. Consequently, possible coordinated activity among granulosa cells (metabolism, macromolecular syntheses, proliferation) may be a response to hormonal stimulation which is mediated by gap junctions (*Motta* and *Van Blerkom*, 1978).

1.3.3. The Zona Pellucida

During the course of oocyte development and follicle growth, an amorphous material, rich in mucopolysaccharides, is deposited in the space between the oocyte and the surrounding granulosa cells. This material is subsequently transformed into a continuous, jelly-like coat – the zona pellucida (Plates 9, 12, 13, 14, 15). The origin of the zona is still somewhat uncertain. Whereas some authors have concluded on the basis of fine structural and histochemical evidence that the zona material is derived from components synthesized by the rough-surfaced endoplasmic reticulum, modified and/or packaged by the Golgi complex and secreted by the granulosa cells (*Trujillo-Cenoz* and *Sotelo*, 1959; *Chiquoine*, 1960), other investigators have suggested that the Golgi vacuoles located in the cortical regions of the oocyte cytoplasm contribute (along with the granulosa cells) to the deposition of the zona pellucida (*Merker*, 1961; *Stegner* and *Wartenberg*, 1961; *Zamboni* and *Mastroianni*, 1966; *Motta, Takeva,* and *Nesci*, 1971; *Martinek* and *Karusova*, 1972; *Baranska, Konwinski,* and *Kujawa*, 1975). Although a rather thick layer (10 to $20 \mu m$) is interposed between the cells of the corona radiata and the surface of the oocyte, intimate contact between these two elements is still maintained. This

contact involves (1) the association of granulosa cell processes and oocyte microvilli, and (2) the establishment of junctions (tight and gap) and junctional complexes between apposing membranes of the oocyte and granulosa cell processes (Plate 15 B) (*Zamboni*, 1974; *Anderson* and *Albertini*, 1976). This relationship usually persists in an unaltered form until ovulation (*Sotelo* and *Porter*, 1959; *Zamboni* and *Mastroianni*, 1966).

As observed by SEM, especially in regions where the oocyte surface is relatively free of granulosa cells, the exterior of the zona has an irregular appearance, which results from the presence of stratified, distorted layers of a dense material (Plates 13, 14 A, 14 B). The zona pellucida of an oocyte in a Graafian follicle generally displays numerous infoldings, canalicular invaginations, crypts, and, at high magnification, a fine coating of granules (Plates 14 B, 15). A number of infoldings appear to be portions of anastomotic channels formed by the evaginations of the granulosa cells that traverse and ramify through the zona to anchor ultimately onto the oocyte surface (Plates 14 C, 15 B). Histochemically, the material that composes the zona is similar to the liquor folliculi that accumulates in the intercellular spaces of preantral follicles.

Granules of the same approximate dimensions as those observed on the surface of the zona by SEM have been demonstrated by TEM to be located within the zona matrix, in close proximity to granulosa cell processes and even in clusters within the processes (Plate 15 B) (*Motta* and *Van Blerkom*, 1974 b). Although several studies support the hypothesis that these granules originate from the cells of the corona radiata (*Zamboni* and *Mastroianni*, 1966; *Pedersen* and *Seidel*, 1972; *Motta* and *Van Blerkom*, 1974 b, 1975), it has been suggested that they arise as small surface protrusions, or "micropapillae," of the plasmalemma of the oocyte and adjacent granulosa cells (at least in the rabbit) (*Pedersen* and *Seidel*, 1972). The composition and function of the granules remain obscure. *Zamboni* and *Mastroianni* (1966) suggest that these granules may be composed of glycogen, but *Pedersen* and *Seidel* (1972) have concluded that the granules are not glycogen. Further study is required in order to elucidate the composition of the granules and their site(s) of origin and also to determine whether they function in the course of oocyte maturation and/or subsequent postovulatory events.

It appears likely that the stratified nature of the zona pellucida in the mature follicle results from a discontinuous secretory process involving both oocyte and granulosa cells. Histochemically, the zona has been shown to be composed of at least two distinct layers with somewhat different chemical compositions (*Wartenberg* and *Stegner*, 1960). The presence of surface granules indicates that, at least in the mature follicle of the rabbit ovary, some "new" material is produced by the granulosa cells directly surrounding the zona, thus indicating a discontinuous process of formation (*Stegner* and *Wartenberg*, 1961; *Motta, Takeva,* and *Nesci,* 1971; *Motta* and *Van Blerkom,* 1974b). However, more detailed biochemical, histological, and fine structural investigation is required in order to resolve fully the question of the origin(s) of the zona pellucida.

1.3.4. Atresia

Many of the morphological transformations of follicle and oocyte that occur during growth are also associated with atretic alteration. As mentioned earlier, not all oocytes and follicles that undergo growth and maturation are necessarily destined to ovulate. The atretic degeneration of an oocyte can occur at *any stage* of follicular development. The number of oocytes that undergo atresia is quite variable and may differ significantly between species. In humans, for example, greater than 99% of all oocytes present at birth undergo atresia at some stage of their development (*Zamboni,* 1972). If atresia occurs in a follicle that is fairly well developed, the follicular components undergo a differentiative process that transforms the granulosa cells of the follicle into lutein-like cells (luteinization; see Chapter 2). The luteinization of an atretic follicle is essentially similar to the luteinization of a normal follicle where, following ovulation, the follicle is transformed into a corpus luteum. However, most atretic follicles undergoing luteinization contain a degenerating oocyte. In some mammals, these special structures occupy interfollicular spaces or large areas of the ovarian stroma, where they appear as irregular corpora lutea (accessory or false corpora lutea). In other cases, the residual structures of atretic follicles take the form of scattered groups of epithelial-like cells derived from theca interna gland cells (see section 1.5); when this occurs, these elements are referred to as interstitial gland cells. The importance of recognizing atretic follicles and oocytes, especially in relation to biochemical and fine structural studies of oocyte maturation *in vivo* and *in vitro* (*McGaughey* and *Van Blerkom,* 1977; *Schultz* and *Wassarman,* 1977a, 1977b; *Van Blerkom,* 1977; *Warnes, Moor,* and *Johnson,* 1977; *Van Blerkom* and *McGaughey,* 1978), has been em-

phasized by several authors (*Motta,* 1972; *Motta, Hadjioloff,* and *Bourneva,* 1972; *Zamboni,* 1972), and their caveats should be carefully considered.

1.4. Theca Interna and Interstitial Cells

During the period of follicle growth, the cells of the stroma that surround the follicle become organized into a series of concentric, compressed layers – the theca folliculi. In younger follicles, the theca folliculi is a sheath of flattened and compressed cells (Plates 9A, 12). With the development of an antrum, the cellular components of the theca folliculi differentiate into two distinct layers:
1. a richly vascular theca interna containing groups of epithelial-like cells (Plates 9B, 16A, 16B, 16C),
2. a theca externa, characterized by the presence of spindle-shaped cells similar in appearance to the cells in the surrounding stroma (Plate 9B).

Gradually, a number of cells from the theca interna enlarge and assume a polyhedral form. The cytoplasm of these cells, which initially contained organelles typical of secretory cells (Golgi complex, RER, ribosomes) (*Santoro,* 1965), now contains elements of the smooth-surfaced endoplasmic reticulum (SER), lipid droplets, and mitochondria with villiform and/or tubular cristae (Plates 16B, 16C). Similar ultrastructural features are characteristic of cells that secrete steroid hormones, and, consequently, the submicroscopic changes occurring within some of the cells of the theca interna have been interpreted as indicating a steroidogenic function (*Christensen* and *Gillim,* 1969). When the above morphological transformations have been completed, the entire structure is referred to as a theca interna gland. Following ovulation, the theca interna glands of some mammals persist for a brief period of time before disappearing entirely (*Mossman* and *Duke,* 1973). However, in other mammals, it is possible that some of the cells of the theca interna differentiate into typical luteal cells (see Chapter 2) (*Motta,* 1936; *Harrison,* 1962).

Closely related in structure to the theca interna glands are the so-called interstitial glands or, more simply, the interstitial cells (Plate 9B). It is commonly believed that the interstitial cells arise from elements of the theca interna located around atretic follicles (*Watzka,* 1957; *Harrison,* 1962; *Mossman* and *Duke,* 1973). Recent histochemical (*Guraya* and *Greenwald,* 1964; *Guraya,* 1967, 1974) and electron microscopic studies (*Motta,* 1966; *Davies* and *Broadus,* 1968) have provided evidence demonstrating the close similarity of interstitial and theca interna cells. Although the more typical origin of interstitial cells may be the theca interna, there is some evidence that in both the immature and mature ovaries of several mammals, interstitial cells, under the appropriate stimulus, may arise directly from the cells of the ovarian stroma (*Merker* and *Diaz-Encinas*, 1969; *Quattropani*, 1973) and/or from isolated groups of epithelial-like cells (*Mori* and *Matsumoto*, 1970; *Motta*, 1974 b). Both interstitial cells and cells of the theca interna are known to undergo morphological transformations that ultimately convey the capability of secreting steroid hormones (*Muta*, 1958; *Motta*, 1966; *Davies* and *Broadus*, 1968; *Mori* and *Matsumoto*, 1970, 1973; *Dahl*, 1971, *Motta*, *Nesci*, and *Fumagalli*, 1971; *Moller*, 1973). At the fine structural level, the cytoplasm of differentiated interstitial gland cells contains the same types of organelles found in steroid-secreting cells – SER, lipid droplets, and mitochondria with tubular and/or villiform cristae (Plate 16 D). In addition, sympathetic nerve fibers have been shown to terminate as large "boutons" embedded in the cytoplasm of differentiated interstitial cells of the mouse (*Unsicker*, 1970) and rabbit ovary (*Motta*, 1974 b). These findings raise the possibility that the secretory activity of differentiated interstitial cells may be under direct neural control! Such control could be an important factor in follicle growth and atresia (*Mossman* and *Duke*, 1973). Other characteristics of interstitial cells that require further study are the nature of the intercellular contact and communication between interstitial cells and the potential role of a pronounced glycocalyx on the cell surface in interstitial cell function (*Motta*, 1974 b; *Familiari*, *Renda*, and *Motta*, 1977).

The chemical composition of the steroid hormone(s) secreted by interstitial and theca interna cells is a continuing subject of debate. In general, histochemical techniques have shown that both theca interna and interstitial gland cells either have the enzymes required for steroidogenesis (*Rubin*, *Deane*, and *Balogh*, 1969; *Motta* and *Bourneva*, 1970) or actually secrete steroid hormones – primarily estrogens (*Hart*, *Baillie*, *Calman*, and *Ferguson*, 1966; *Albrecht*, *Koos*, and *Wehrenberg*, 1975; *Motta* and *Takeva*, 1971). One major exception to this set of observations is that interstitial cells in the rabbit ovary can effectively substitute for a corpus luteum in the production of progesterone (*Hilliard*, *Hayward*, and *Sawyer*, 1964). If differentiated interstitial cells secrete primarily estrogens (or androgens), then one possible regulative function of these ovarian elements could be

in the development and maintenance of the secondary female sexual characteristics, just as in the male the secretion of testicular interstitial cells regulates male sexual characteristics (*Mossman* and *Duke*, 1973).

1.5. The Ovarian Stroma and Smooth Muscle Cells

The ovarian stroma is composed of a special type of poorly differentiated, embryonal-mesenchymal-like cell capable of undergoing a series of complex morphological alterations during the reproductive life of the animal (Plate 17). As a result of these alterations, stromal cells can give rise to the following: theca interna cells, interstitial cells, and possibly, on occasion, granulosa cells, "theca-lutein," or "paraluteal" cells (*Mossman* and *Duke*, 1973). Stromal cells that do not undergo direct differentiation are believed to be capable of altering their morphology in accordance with their location within the ovary and with the phase of the estrous/menstrual cycle (*Balboni*, 1970).

A matter of considerable debate has concerned the presence of smooth muscle cells (also known as myoid cells) in the mammalian ovary (*Motta*, 1929; *Claesson*, 1947; *Watzka*, 1957). On the basis of numerous TEM studies, it has been claimed that smooth muscle cells are present throughout the ovary and especially in the cortical stroma, where they intermingle with the cells of the theca (*Fumagalli* and *Motta*, 1969; *O'Shea*, 1970; *Osvaldo-Decima*, 1970; *Fumagalli*, *Motta*, and *Calvieri*, 1971; *Burden*, 1972; *McReynolds*, *Siraki*, *Bramson*, and *Pollock*, 1973). Myoid cells appear to be similar to smooth muscle cells in other tissues insofar as they contain large bundles of microfilaments, micropinocytotic vesicles located just beneath the plasma membrane (Plate 17 C) and deeply infolded or indented nuclei. However, the potential for mistaking myoid cells for other ovarian elements containing microfilaments (e.g., fibroblasts) exists, and therefore, strict fine structural and biochemical criteria should be applied to their identification, especially in the perifollicular regions of the ovary. Ultrastructural and histochemical evidence suggests that myoid cells lie in proximity to, or are in contact with, cholinergic nerve terminals such that potential neuromuscular junctions may be formed (*Bahr*, *Kao*, and *Nalbandov*, 1974; *Walles*, *Edvinsson*, *Owman*, *Sjoberg*, and *Sporrong*, 1976). In addition, *Walles* and his collaborators (1976) have indicated the presence of cholinergic receptors on what they consider to be myoid cells and suggest that such receptors may mediate the contraction of the bovine Graafian follicle.

References

Adams, E. G., and Hertig, A. T. (1964). Studies on guinea pig oocytes. I. Electron microscopic observations on the development of cytoplasmic organelles in oocytes of primordial and primary follicles. J. Cell Biol. 21:397–427.

Albertini, D. F., and Anderson, E. (1974). The appearance and structure of intercellular connections during the ontogeny of rabbit ovarian follicles with particular reference to gap junctions. J. Cell Biol. 63:234–250.

Albrecht, E. D., Koos, R. D., and Wehrenberg, W. B. (1975). Ovarian Δ^5-3-β-hydroxysteroid dehydrogenase and cholesterol in the aged mouse during pregnancy. Biol. Repr. 13:158–162.

Anderson, E. (1971). Intercellular junctions in the differentiating Graafian follicle of the mouse. Anat. Rec. 169:473 (abstract).

Anderson, E., and Albertini, D. F. (1976). Gap junctions between the oocyte and companion follicle cells in the mammalian ovary. J. Cell Biol. 71:680–686.

Anderson, E., Lee, G., Letourneau, R., Albertini, D. F., and Meller, S. M. (1976). Cytological observations of the ovarian epithelium in mammals during the reproductive cycle. J. Morph. 150:135–166.

Andrews, P. M., and Porter, K. R. (1973). The ultrastructural morphology and possible functional significance of mesothelial microvilli. Anat. Rec. 177:409–426.

Ansalone, G., Familiari, G., Fedele, F., and Motta, P. (1973). Sulla presenza di microfilamenti nel corso della differenziazione delle cellule della granulosa in cellule luteiniche nel coniglio. Boll. Soc. Ital. Biol. Sper. 49:561–565.

Baca, M., and Zamboni, L. (1967). The fine structure of human follicular oocytes. J. Ultrastruct. Res. 19:354–381.

Bahr, J., Kao, L., and Nalbandov, A. V. (1974). The role of catecholamines and nerves in ovulation. Biol. Reprod. 10:273–290.

Baker, T. G., and Franchi, L. L. (1967). The fine structure of oogonia and oocytes in human ovaries. J. Cell Sci. 2:213–224.

Baker, T. G., Beaumont, H. M., and Franchi, L. L. (1969). The uptake of tritiated uridine and phenylalanine by the ovaries of rats and monkeys. J. Cell Sci. 4:655–675.

Balboni, G. C. (1970). Le strome de l'ovarie humain. Aspects ultrastructuraux. Bull. Ass. Anat. Francais 147:105–110.

Balboni, G. C. (1976). Histology of the ovary. In: The Endocrine Function of the Human Ovary (V. H. T. James, M. Serio, and G. Giusti, eds.). Proc. Serono Symposia, Vol. 7, pp. 1–24. Academic Press, New York.

Baranska, W. Konwinski, M., and Kujawa, M. (1975). Fine structure of the zona pellucida of unfertilized egg cells and embryos. J. Exp. Zool. 192:193–202.

Bjorkman, N. (1962). A study of the ultrastructure of granulosa cells of the rat ovary. Acta Anat. (Basel) 51:125–147.

Blanchette, E. J. (1961). A study of the fine structure of the rabbit primary oocyte. J. Ultrastruct. Res. 5:349–363.

Bouin P. and Ancel P. (1908). Sur la différenciation d'une membrane propre d'origine épithéliale pendant le dévelopment du corps jaune chez le chienne. C. R. Soc. Biol., Paris 65:201–235.

Brambell, F. W. R. (1928). The development and morphology of the gonads of the mouse. III. The growth of the follicle. Proc. R. Soc. B. 103:258–271.

Burden, H. W. (1972). Ultrastructural observations on ovarian perifollicular smooth muscle in the cat, guinea pig and rabbit. Amer. J. Anat. 133:125–142.

Byskov, A. G. (1969). Ultrastructural studies on the preovulatory follicle in the mouse ovary. Z. Zellforsch. mikrosk. Anat. 100:285–299.

Call, E., and Exner, S. (1875). Zur Kenntnis des Graafschen Follikels und des Corpus luteum beim Kaninchen. S.-B. Akad. Wiss. Wien. 71:321–328.

Cavallotti, C., DiDio, L. J. A., Familiari, G. Fumagalli, G., and Motta, P. (1975). Microfilaments in granulosa cells of rabbit ovary: Immunological and ultrastructural observations. Acta Histochemica 52:253–256.

Cherney, D. D., Motta, P., and DiDio, L. J. A. (1973). Ovarian villi in rabbits studied with light scanning and transmission electron microscopy. J. Microscopie 17:37–40.

Chiquoine, A. D. (1960). The development of the zona pellucida of the mammalian ovum. Amer. J. Anat. 106:149–170.

Christensen, A. K., and Gillim, S. W. (1969). The correlation of fine structure and function in steroid secreting cells with emphasis on those of the gonads. In: The Gonads, pp. 415–488. (McKerns, K. S. ed.). Appleton-Century, Crofts, New York.

Claesson, L. (1947). Is there any smooth muscle in the wall of the Graafian follicle? Anat. (Basel) 3:295–311.

Dabelow, A. (1939). Das Gefäßnetz des Ovars und sein Verhalten während der zyklischen Veränderungen. Anat. Anz. 88:173–182.

Dahl, E. (1971). Studies of the fine structure of ovarian interstitial tissue. I. A comparative study of the fine structure of the ovarian interstitial tissue in the rat and the domestic fowl. J. Anat. (London) 108:275–290.

Dalcq, A. M. (1963). The relation to lysosomes of the in vivo metachromatic granules. In: Ciba Foundation Symposium on Lysosomes (de Reuck, A. V. S. and Cameron, M. P. eds.). Little, Brown & Co., Boston.

Davies J., and Broadus, C. D. (1968). Studies on the fine structure of ovarian steroid secreting cells in the rabbit. I. The normal interstitial cells. Amer. J. Anat. 123:441–474.

Eddy, E. M. (1974). Fine structural observations on the form and distribution of "nuage" in germ cells of the rat. Anat. Rec. 178:731–758.

Espey, L. L., and Stutts, R. H. (1972). Exchange of cytoplasm between cells of the membrana granulosa in rabbit ovarian follicles. Biol. Reprod. 6:168–175.

Familiari, G., Renda, T., and Motta, P. (1977). A surface coat in steroid secreting cells of the mouse ovary. Acta Anat. (Basel), 100:193–202.

Ferenczy, A., and Richart, R. M. (1974). Female reproductive system. Dynamic of scan and transmission electron microscopy. John Wiley & Sons, New York.

Franceschini, M. P., Santoro, A., and Motta, P. (1965). L'ultrastruttura delle cellule della granulosa nelle varie fasi di maturazione del follicolo ooforo. Z. Anat. Entw. Gesch. 124:522–532.

Franchi, L. L. (1960). Electron microscopy of oocyte-folli-cle cell relationship in the rat ovary. J. Biophys. Biochem. Cytol. 7:397–398.

Fumagalli, Z., and Motta, P. (1969). Sulla presenza al microscopio elettronico di cellule muscolari lisce nell'ovaio di alcuni mammiferi. Atti. Soc. Ital. Anat. 28th Conv. Soc., Napoli.

Fumagalli, Z., Motta, P., and Calvieri, S. (1971). The presence of smooth muscle cells in the ovary of several mammals as seen under the electron microscope. Experientia 27:682–683.

Greep, R. P. (1963). Histology, histochemistry and ultra-structure of the adult ovary, In: The Ovary, p. 48 (Grady, H. G., and Smith, D. E., eds.). Acad. Pathol. Monogr. No. 3. Williams and Wilkins Co., Baltimore, Md.

Guraya, S. (1967). Histochemical study of the interstitial gland tissue in the ovaries of nonpregnant women. Amer. J. Obstet. Gynec. 98:99–106.

Guraya, S. (1974). Comparative morphological and his-tochemical observations on the ovarian stromal compartment in mammals with special reference to steroidogenesis. Acta. Anat. (Basel) 90:250–284.

Guraya, S. and Greenwald, G. S. (1964). Histochemical studies on the interstitial gland in the rabbit ovary. Amer. J. Anat. 114:495–519.

Hadek, R. (1963). Electron microscope study on primary liquor folliculi secretion in the mouse ovary. J. Ultrastruct. Res. 9:445–458.

Hadek, R. (1969). Mammalian fertilization. An Atlas of Ultrastructure. Academic Press, New York.

Harrison, R. J. (1962). Ovarian structure (mammals). In: The Ovary, Vol. 1. pp. 143–187, (Zuckerman, S., Mandl. A. M., and Eckstein, P., eds.). Academic Press, New York.

Harrison, R. J., and Matthews, L. H. (1951). Subsurface crypts in the cortex of the mammalian ovary. Proc. Zool. Soc. London 120:699–712.

Hart, D. M., Baillie, A. H., Calman, K. C., and Ferguson, M. M. (1966). Hydroxysteroid dehydrogenase development in the mouse adrenals and gonads. The Ovary. In: Development in Steroid Histochemistry, Chapter 5. (Baillie, A. H., Ferguson, M. M., and Hart, D. McK., eds.). Academic Press, New York.

Hertig, A. T. (1968). The primary human oocyte. Some observations of the fine structure of Balbiani's vitelline body and the origin of the annulate lamellae. Amer. J. Anat. 122:107–138.

Hertig, A. T., and Adams, E. C. (1967). Studies on the human oocyte and its follicle. Ultrastructural and cytochemical observations on the primordial follicle stage. J. Cell Biol. 34:647–675.

Hilliard, J., Hayward, J. N., and Sawyer, C. H. (1964). Post coital patterns of secretion of pituitary gonadotropin and ovarian progestin in the rabbit. Endocrinology 75:957–963.

Hope, J. (1965). The fine structure of the developing follicle of the Rhesus ovary. J. Ultrastruct. Res. 12:592–610.

Jacoby, A. (1962). Histochemistry. In: The Ovary, Vol. 1 (Zuckermans, Mandl, A. M., and Eckstein, P., eds.). Academic Press, New York.

Jensen, R. D, and Norris, H. J. (1972). Epithelial tumors of the ovary. Arch. Path. 94:29–34.

Jeppesen, T. (1975). Surface epithelium of the fetal guinea pig ovary. A light and electron microscopic study. Anat. Rec. 183:499–516.

Krarup, T., Pedersen, T., and Faber, M. (1969). Regulation of oocyte growth in the mouse ovary. Nature (London) 224:187–188.

Ludwig, H., and Metzger, H. (1976). The human female reproductive tract. A scanning electron microscopic atlas. Springer-Verlag, Berlin–Heidelberg–New York.

Mahowald, A. P. (1962). Fine structure of pole cells and polar granules in Drosophila melanogaster. J. Exp. Zool. 151:201–215.

Martinek, J., and Karusova, H. (1972). Development of the zona pellucida in the rat. Fol. Morphol. 20:73–75.

McGaughey, R. W., and Van Blerkom, J. (1977). Patterns of polypeptide synthesis of porcine oocytes during maturation in vitro. Devel. Biol. 56:241–254.

McReynolds, H. D., Siraki, C. M., Bramson, P. H., and Pollock, R. J. (1973). Smooth muscle-like cells in the ovaries of the hamster and gerbil. Z. Zellforsch. mikrosk. Anat. 140:1–8.

Merk, F. B., Albright, J. T., and Botticelli, C. R. (1973). The fine structure of granulosa cell nexuses in rat ovarian follicles. Anat. Rec. 175:107–125.

Merker, H. J. (1961). Elektronenmikroskopische Untersuchungen über die Bildung der Zona Pellucida in den Follikeln des Kaninchenovars. Z. Zellforsch. mikrosk. Anat. 54:677–688.

Merker, H. J., and Diaz-Encinas, J. (1969). Das electronenmikroskopische Bild des Ovars juveniler Ratten und Kaninchen nach Stimulierung mit PMS und HCG. I. Theka und Stroma (Interstitiellen Drüse). Z. Zellforsch. mikrosk. Anat. 94:605–623.

Mestwerdt, W., Müller, O., and Brandau, H. (1977). Die differenzierte Struktur und Function der Granulosa und Theka in verschieden Follikelstadien menschlicher Ovarien. I. Mitteilung: Der Primordial Primar-, Sekundar- und ruhende Tertiarfollikel. Arch. Gynak. 45–71.

Moller, O. M. (1973). The fine structure of the ovarian interstitial gland cells in the mink, Mustela vison. J. Reprod. Fertil. 34:171–174.

Moore, G. P. M., Lintern-Moore, S., Peters, H., and Faber, M. (1974). RNA synthesis in the mouse oocyte. J. Cell Biol. 60:416–422.

Mori, H., and Matsumoto, K. (1970). On the histogenesis of the ovarian interstitial gland in the rabbit. I. Primary interstitial gland. Amer. J. Anat. 129:289–306.

Mori, H., and Matsumoto, K. (1973). Development of the secondary interstitial gland in the rabbit ovary. J. Anat. (London) 116:417–430.

Mossman, H. W., and Duke, K. L. (1973). Comparative Morphology of the Mammalian Ovary. University of Wisconsin Press, Madison, Wisc.

Motta, G. (1929). Sull' importanza delle cellule muscolari dell'ovaio e sul meccanismo della deiscenza del follicolo. Riv. Ital. Ginecol. 10:1–56.

Motta, G. (1936). Beobachtungen und Betrachtungen über die Genese des Corpus luteum der Granulosa des Follikels. Zentr. Gynak. 26:1547–1555.

Motta, P. (1965 a). Sulla presenza al microscopio elettronico di ciglia e centrioli in cellule della granulosa ovarica di Lepus cuniculus (Linn.). Boll. Soc. Ital. Biol. sper. 41:31–35.

Motta, P. (1965 b). Sur l'ultrastructure des "Corps des Call et Exner" dan l'ovarie du lapin. Z. Zellforsch. mikrosk. Anat. *68*:308–319.

Motta, P. (1966). Osservazioni sulla fine struttura delle cellule interstiziali dell'ovaio. Biol. lat. *19*:107–137.

Motta, P. (1972). Histochemical evidence of early stages of atretic follicles in different mammals. In: Proceedings of the 4th International Congress on Histochemistry and Cytochemistry, pp. 599–600 (Takeuchi, Ogawa, and Fujita, eds.). Kyoto.

Motta, P. (1974a). The fine structure of ovarian cortical crypts and cords in mature rabbits. A transmission and scanning electron microscopic study. Acta Anat. (Basel) *90*:36–64.

Motta, P. (1974b). Superficial epithelium and surface evaginations in the cortex of mature rabbit ovaries. A note on the histogenesis of the interstitial cells. Fert. Steril. *25*:336–347.

Motta, P. and Nesci, E. (1969). The Call and Exner bodies of mammalian ovaries with reference to the problem of rosette formation. Arch. Anat. Micr. Morph. Exp. *58*:283–290.

Motta, P., and Bourneva, V. (1970). A comparative histochemical study of Δ^5-3β-OHD and lipid content in the rat ovary with special reference to the interstitial cells. Acta Histochem. *38*:340–351.

Motta, P., Cherney, D. D., and DiDio, L. J. A. (1971). Scanning and transmission electron microscopy of the ovarian surface in mammals with special reference to ovulation. J. Submicr. Cytol. *3*:85–100.

Motta, P., Nesci, E., and Fumagalli, L. (1971). The fine structure and cyclic morphological changes of the interstitial cells in the mammalian ovary. Arc. Hist. Embr. Norm. et Exp. *54*:43–58.

Motta, P., and Takeva, Z. (1971). Histochemical demonstration of hydroxysteroid dehydrogenase activity in the interstitial tissue of the guinea pig ovary during the oestrus cycle and pregnancy. Fert. Steril. *22*:378–382.

Motta, P., Takeva, Z., and Nesci, E. (1971). Etude ultrastructurale et histochimique des rapports entre les cellules folliculaires et l'ovocyte pendant le development du follicle ovarien chez le mammifers. Acta Anat. (Basel). *80*:537–562.

Motta, P., Takeva, Z., and Palermo, D. (1971). On the presence of cilia in different cells of the mammalian ovary. Acta Anat. (Basel) *78*:591–603.

Motta, P., Hadjioloff, A. I., and Bourneva, V. (1972). Über die Ultrastruktur und die Δ^5-3-β-Hydroxysteroid-Dehydrogenase-Aktivität der Granulosazellen im Eierstock der Ratte und Maus im Laufe der Postnatalentwicklung. Anat. Anz. *130*:615–623.

Motta, P. and DiDio, L. J. A. (1974). Microfilaments in granulosa cells during the development of the follicle and its transformation in corpus luteum in the rabbit ovary. J. Submicr. Cytol. *6*:15–27.

Motta, P., and Van Blerkom, J. (1974a). Presence d'un material characteristique granulair dans le cytoplasme de l'ovocyte et dans les premiers stades de la differenciation des cellules embryonnaires. Bull. Ass. Anat. (Francais) *58*:947–953.

Motta, P., and Van Blerkom, J. (1974b). A scanning electron microscopie study of the luteo-follicular complex. I. Follicle and oocyte. J. Submicr. Cytol. *6*:297–310.

Motta, P., and Van Blerkom, J. (1975). A scanning electron microscopic study of the luteo-follicular complex. II. Events leading to ovulation. Amer. J. Anat. *143*:241–264.

Motta, P., Andrews, P. M., and Porter, K. R. (1977). Microanatomy of Cells and Tissue Surfaces: An Atlas of Scanning Electron Microscopy. Vallardi and Lea-Febigers, Milano and Philadelphia.

Motta, P., and Van Blerkom, J. (1978). Structure and ultrastructure of the Graafian follicle. In: Human Ovulation: Mechanisms, Prediction, Detection and Regulation (E. S. E. Hafez, ed.). North Holland, Amsterdam.

Muta, T. (1958). The fine structure of the interstitial cell in the mouse ovary studied with the electron microscope. Kurume Med. J. *5*:167–185.

Nicosia, S. V., and Mikhail, G. (1975). Cumuli oophori in tissue culture: hormone production, ultrastructure and morphometry of early luteinization. Fert. Steril. *26*:427–448.

Oakberg, E. F. (1968). Relationship between stage of follicle development and RNA synthesis in the mouse oocyte. Mutat. Res. *6*:155–165.

Odor, D. L. (1960). Electron microscopic studies on ovarian oocytes and unfertilized tubal ova in the rat. J. Biophys. Biochem. Cytol. *7*:567–574.

Odor, D. L. (1965). The ultrastructure of unilaminar follicles of the hamster ovary. Amer. J. Anat. *116*:493–522.

O'Shea, J. D. (1970). An ultrastructural study of smooth muscle-like cells in the theca externa of ovarian follicles in the rat. Anat. Rec. *167*:127–140.

Osvaldo-Decima, L. (1970). Smooth muscle in the ovary of the rat and monkey. J. Ultrastruct. Res. *30*:218–237.

Papadaki, L., and Belby, J. O. W. (1971). The fine structure of the surface epithelium of human ovary. J. Cell Sci. *8*:445–465.

Pavelka, R., Friedrich, F., and Caucig, H. (1972). Die Ultrastructur der menschlichen Eizelle des Graafschen Follikels. Wien Klin. Wochenschr. *84*:305–312.

Pedersen, T. (1969). Follicle growth in the immature mouse ovary. Acta endocr. (Copenhagen) *62*:117–132.

Pedersen, T., and Peters, H. (1968). Proposal for the classification of oocytes and follicles in the mouse ovary. J. Reprod. Fert. *17*:555–557.

Pedersen, H., and Seidel, G. (1972). Micropapillae: A local modification of the cell surface observed in rabbit oocytes and adjacent follicle cells. J. Ultrastruct. Res. *39*:540–548.

Peters, H. (1969). The development of the mouse ovary from birth to maturity. Acta endocr. (Copenhagen) *62*:98–116.

Peters, H., Byskov, A. G., Lintern-Moore, S., Faber, M., and Andersen, M. (1973). The effect of gonadotrophin on follicle growth initiation in the neonatal mouse ovary. J. Reprod. Ferr. *35*:139–141.

Porter, K. R., Prescott, D., and Frye, J. (1973). Changes in the surface morphology of Chinese hamster ovary cells during the cell cycle. J. Cell Biol. *57*:815–836.

Quattropani, S. (1973). Morphogenesis of the ovarian interstitial tissue in the neonatal mouse. Anat. Rec. *177*:569–584.

Raven, C. P. (1961). Oogenesis. The Storage of Developmental Information. Pergamon Press, New York.

Rodman, T. C., and Bachvarova, R. (1976). RNA synthesis in preovulatory mouse oocytes. J. Cell Biol. *70*:251–257.

Rubin, B. L., Deane, H. W., and Balogh, K. (1969). Ovarian

steroid biosynthesis and Δ^5-3-β and 20 -hydroxysteroid dehydrogenase activity. Trans. N. Y. Acad. Sci. Ser. II. *31*:787–802.

Santoro, A. (1965). On the fine structure of the theca interna cells of the rabbit ovary. Boll. Soc. Ital. Biol. Sper. *40*:1636–1637.

Sato, S. (1965). An electron microscope study of the fine structure of the ovary in normal mature rats. Arc. Histol. Jap. *26*:115–149.

Schuchner, E. B., and Stockert, J. C. (1973). Filaments and microtubules in the cytoplasm of the granulosa cells from the human follicle. Protoplasma 76:133–137.

Schultz, R. M., and Wassarman, P. M. (1977 a). Biochemical studies of mammalian oogenesis: Protein synthesis during oocyte growth and meiotic maturation in the mouse. J. Cell Sci. *24*:167–194.

Schultz, R. M., and Wassarman, P. M. (1977 b). Specific changes in the pattern of protein synthesis during meiotic maturation of mammalian oocytes *in vitro*. Proc. Natl. Acad. Sci. (USA) 74:538–541.

Smith, L. D., and Williams, M. A. (1975). Germinal plasm and determination of primordial germ cells. In: The Developmental Biology of Reproduction (Market, C. L. and Papaconstantinou, J. eds.). Academic Press, New York.

Sotelo, J. R. and Porter, K. R. (1959). An electron microscope study of the rat ovum J. Biophys. Biochem. Cytol. 5:327–341.

Stegner, H. E. (1967). Die electronmikroskopische Struktur der Eizelle. Ergeb. Anat. Entwicklungsgesch. *39*:1–112.

Stegner, H. E., and Wartenberg, H. (1961). Electronmikroskopische und histopochemische Untersuchungen über Struktur und Bildung der Zona pellucida Menschlicher Eizelle. Z. Zellforsch. mikrosk. Anat. 53:702–713.

Sternberg, W. H. (1963). Non-functioning ovarian neoplasm. In: The Ovary (Grady H. S. and Smith, D. E. eds.). Williams and Wilkins Co., Baltimore, Md.

Szollosi, D. (1967). Development of cortical granules and the cortical reaction in rat and hamster eggs. Anat. Rec. *159*:431–466.

Tanaka, K., Sugawara, S., and Hafez, E. S. E. (1975). The mammalian ovary. In: Scanning Electron Microscopic Atlas of Mammalian Reproduction, pp. 112–127, (E. S. E. Hafez, ed.). Igaku-Shoin, Ltd., Tokyo.

Trujillo-Cenoz., and Sotelo, J. R. (1959). Relationship of the ovular surface with follicle cells and the origin of the zona pellucida in rabbit oocytes. J. Biophys. Biochem. Cytol. 5:347–350.

Unsicker, K. (1970). Zur Innervation der interstitiellen Drüse im Ovar der Maus (Mus musculus). Eine fluoreszent und electronmikroskopische Studie. Z. Zellforsch. mikrosk. Anat. *109*:46–54.

Unsicker, K. (1971). Über den Feinbau von Marksträngen und Markschläuchen im Ovar juveniler und geschlechtsreifer Schweine (Sus scrofa, L.). Z. Zellforsch. mikrosk. Anat. *114*:344–364.

Van Blerkom, J., and Manes, C. (1974). Development of preimplantation rabbit embryos *in vivo* and *in vitro*. II. A comparison of qualitative aspects of protein synthesis. Devel. Biol. 40:40–51.

Van Blerkom, J., and Brockway, G. O. (1975). Qualitative patterns of protein synthesis in the preimplantation mouse embryo. I. Normal pregnancy. Devel. Biol. 44:148–157.

Van Blerkom, J. (1977). Molecular approaches to the study of oocyte maturation and preimplantation embryonic development. In: Immunobiology of the Gametes (Edidin, M. and Johnson, M. H., eds.), pp. 187–206. Cambridge University Press, Cambridge, England.

Van Blerkom, J., and McGaughey, R. W. (1978). Molecular differentiation of the rabbit ovum. I. During oocyte maturation *in vivo* and *in vitro*. Develop. Biol. 63:139–150.

Walles, B., Edvinsson, L., Owman, C., Sjoberg, N.-O., and Sporrong, B. (1976). Cholingeric nerves and receptors mediating contraction of the Graafian follicle. Biol. Reprod. 15:565–572.

Warnes, G. M., Moor, R. M., and Johnson, M. H. (1977). Changes in protein synthesis during maturation of sheep oocytes *in vivo* and *in vitro*. J. Reprod. Fert. 49:331–335.

Wartenberg, H., and Stegner, H. E. (1960). Über die elektronenmikroskopische Feinstruktur des Menschlichen Ovarialeis. Z. Zellforsch. mikrosk. Anat. 52:450–474.

Wassarman, P. M., and Letourneau, G. E. (1976). RNA synthesis in fully grown mouse oocytes. Nature (London) 261:73–74.

Watzka, A. M. (1957). Weibliche Genitalorgane. Das Ovarium. In: Handbuch der Mikroskopischen Anatomie des Menschen, Vol. 7, pp. 1–178 (Mollendorf and Bargmann, eds.). Springer-Verlag, Berlin.

Weakley, B. S. (1966). Electron microscopy of the oocyte and granulosa cells in the developing ovarian follicles of the golden hamster (Mesocricetus auratus). J. Anat. (London) *100*:503–534.

Weakley, B. S. (1967). "Balbiani body" in the oocyte of the golden hamster. Z. Zellforsch. mikrosk. Anat. 83:582–588.

Weakley, B. S. (1969). Differentiation of the surface epithelium of the hamster ovary. An electron microscopic study. J. Anat. (London) *105*:129–147.

Wischnitzer, S. (1965). The ultrastructure of the germinal epithelium of the ovary. J. Morph. *117*:387–400.

Yamada, E., Muta, T., Motomura, A., and Koga, H. (1957). The fine structure of the oocyte in the mouse studied with the electron microscope. Kurume Med. J. *4*:148–171.

Zamboni, L. (1970). Ultrastructure of mammalian oocytes and ova. Biol. Reprod. 2, suppl. 2:44–63.

Zamboni, L. (1971). Fine Morphology of Mammalian Fertilization. Harper & Row, New York.

Zamboni, L. (1972). Comparative studies on the ultrastructure of mammalian oocytes. In: Oogenesis (Biggers, J. D. and Schuetz, A. W. eds.). University Park Press, Baltimore, Md.

Zamboni, L. (1974). Fine morphology of the follicle wall and the follicle cell-oocyte association. Biol. Reprod. *10*:125–149.

Zamboni, L., and Mastroianni, L. (1966). Electron microscopie studies on rabbit ova. I. The follicular oocyte. J. Ultrastruct. Res. *14*:95–117.

Zuckerman, S. and Weir, B. J. (1977). The Ovary Vol. 1, 2nd ed. Academic Press, New York.

Plate 1. The Superficial Epithelium of the Ovary.

A   A general view of the superficial epithelium of the ovary. Most cells are polygonal and approximately equal in size. However, some cells are obviously larger and likely represent cells close to mitosis (arrows). (x 300; estrous rabbit).

B   At high magnification, it is clear that the surfaces of the superficial cells contain numerous microvilli, blebs, and ruffles of the plasma membrane. (x 4,800; estrous rabbit).

C   The appearance of blebs and ruffles is more evident at a still higher magnification. (x 7,500; estrous rabbit).

D   This scanning electron micrograph demonstrates the typical appearance of microvilli on the superficial epithelium. (7,500; proestrous rat).

E   High-magnification scanning electron micrograph of the surface of a superficial cell. Among the microvilli a single cilium is quite evident (arrow). The cilium is partially invaginated into the cortical cytoplasm. Compare this aspect of the cilium as observed by SEM with the cilium illustrated in Plate 7 D (TEM). The surfaces of the superficial cells not covered by microvilli display numerous invaginations or "pits" which open into cortical caveolae (small white arrows). (x 28,000; estrous rabbit).

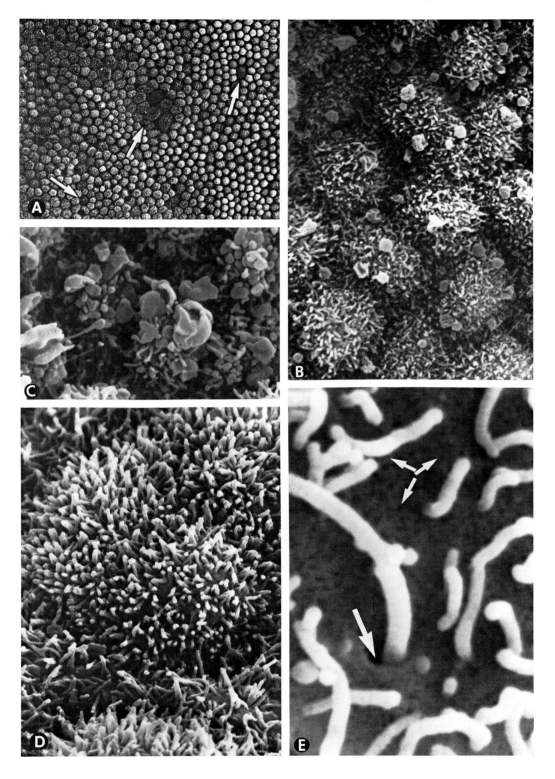

Plate 1. The Superficial Epithelium of the Ovary.

Plate 2. Ovarian Cortical Crypts and Cords.

A   Transmission electron micrograph of the ovarian cortex. The superficial epithelium (Se) penetrates into the tunica albuginea (Ta) and in the process forms cords (C) and crypts (Cy) which were formerly thought to contribute to the oocyte population. The fine structure of the cells composing the crypts and cords is quite similar in appearance to the cells of the superficial epithelium. (x 2,400; estrous rabbit).

B   Transmission electron micrograph of a "nest" of cord cells. These elements possess an irregular, infolded nucleus (N), lipid droplets (L), numerous free ribosomes, polysomes, and a few membranes of the endoplasmic reticulum. Also evident within the cytoplasm is a Golgi complex (G) and occasional lysosomes (Ly). A nerve terminal appears to come into close contact with a cord cell (arrow). Cord cells are very similar in fine structure to follicle cells in adjacent, developing follicles, and also to cells of the superficial epithelium. Compare this figure with Plates 7D, 11B, and 12. (x 4,575; estrous rabbit).

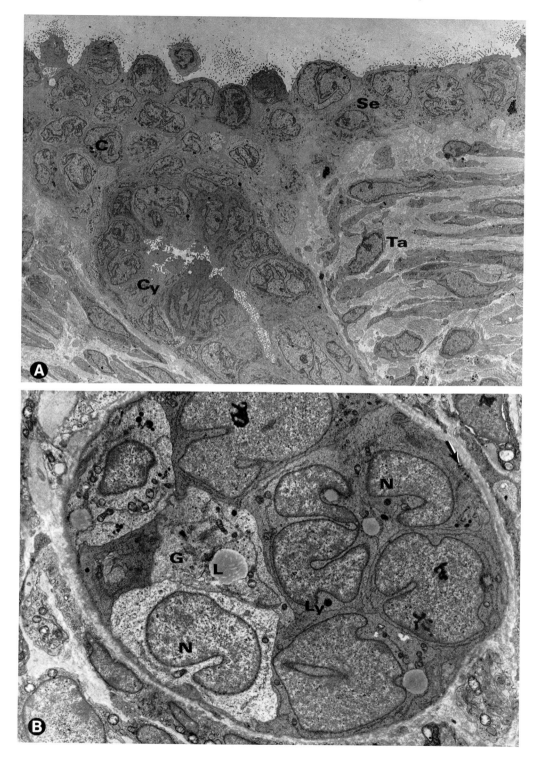

Plate 2. Ovarian Cortical Crypts and Cords.

Plate 3. Ovarian Papillae (Rabbit).

A   Numerous villous-like projections, or "papillae," arise from the cortex of the ovary by a process of evagination and are illustrated in this scanning electron micrograph. (x 70; estrous rabbit).

B   Several relatively small evaginations or "papillae" (arrows) surround an area of the ovarian surface which also contains a comparatively enormous papilla (P). (x 340; rabbit on day 5 postcoitum).

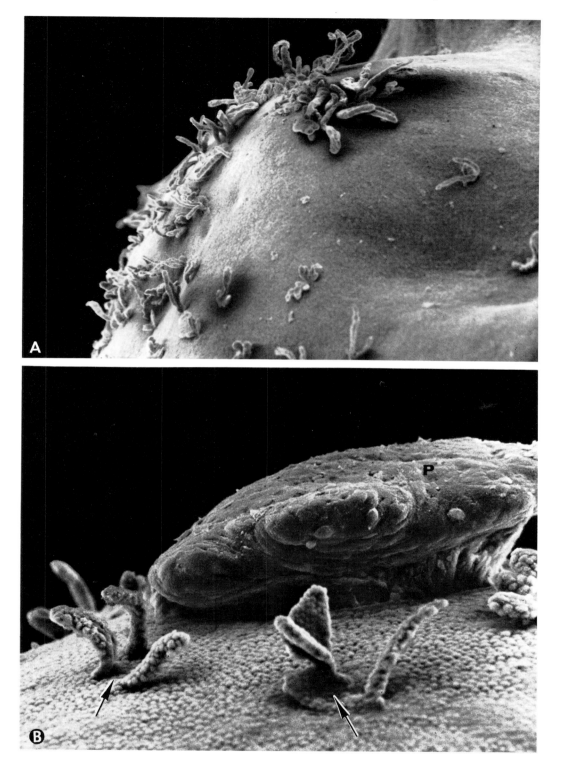

Plate 3. Ovarian Papillae (Rabbit).

Plate 4. Ovarian Papillae (Rat).

A   Some of the superficial cells of the ovary evaginate and, in the process, form small groups of protruding structures, or papillae. If compared to the superficial appearance of adjacent cells, the cells composing this papilla have a rather smooth surface and fewer microvilli (arrows). On the contrary, the microvilli (Mv) are numerous on the surfaces of adjacent cells. The smooth surface and spherical shape of these cells are likely a reflection of the phase of the cell cycle; these particular cells are probably close to mitosis (x 3,280; proestrous rat).

B   This scanning electron micrograph illustrates the appearance of a typical small papilla. Papillae in the mouse and rat never attain the degree of development and relative enormity of size observed in the rabbit. In this particular papilla, some cells display numerous microvilli (Mv) and blebs (B), while others possess rather smooth surfaces (arrow). The superficial aspects of these cells are most likely related to the phase of the cell cycle. (x 3,190; proestrous rat).

24

Plate 4. Ovarian Papillae (Rat).

Plate 5. Ovarian Crypts.

A   Even at relatively low magnification, numerous crater-like invaginations, or crypts, are clearly evident on the surface of the ovary. While some crypts are circular, others have a more elongated shape (arrows). (x 220; estrous rabbit).

B   A higher magnification view of an ovarian crypt and the cells surrounding it. Several different morphologies are displayed by the cells surrounding crypts: some cells possess numerous microvilli (Mv), whereas others have a smooth surface (*) and appear either elongated or flattened. (x 2,450; estrous rat).

C   This scanning electron micrograph illustrates the rather characteristic appearance of those cells located near the luminal surface of a crypt. Typically, these cells contain few microvilli but have an irregular appearance due to the presence of numerous blebs (B). (x 5,400; estrous rabbit).

D   Some of the cells which border the lumen of an ovarian crypt are highly flattened, irregular in shape, and possess few, short microvilli (Mv). (x 6,000; estrous rabbit).

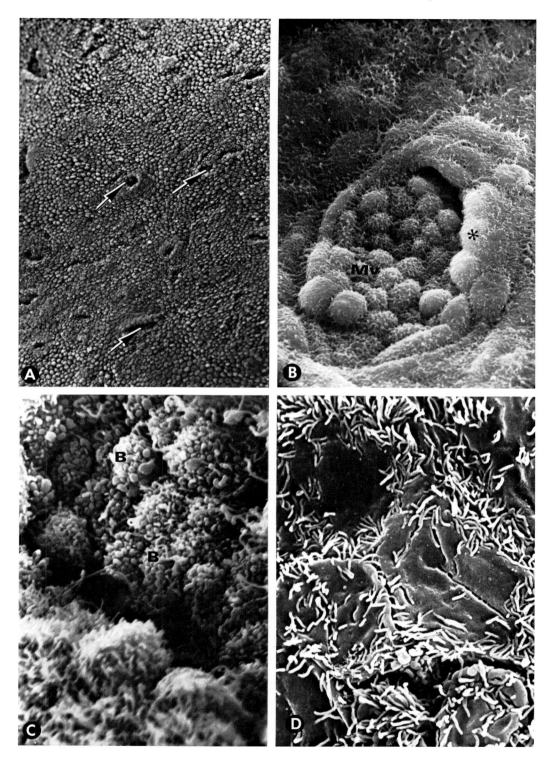

Plate 5. Ovarian Crypts.

Plate 6. Stereo View of Ovarian Papillae and Crypts.

A   Papillae of various sizes are evident in this region of the ovarian surface. The larger papilla seems to have been formed by the coalescence of smaller, adjoining papillae. (x 312; rabbit on day 6 postcoitum).

B   This particular crypt invaginates quite deeply into the ovarian cortex. Whereas the cells of the superficial epithelium possess numerous microvilli, the cells which immediately surround the crypt or line the lumen display either a reduced population of microvilli and a relatively smooth surface or a surface containing a large number of blebs. (x 2,200; estrous rabbit).

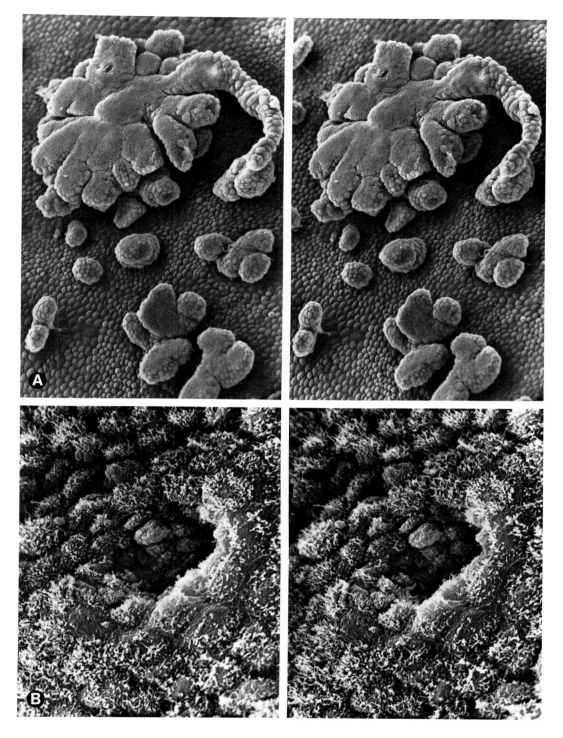

Plate 6. Stereo View of Ovarian Papillae and Crypts.

Plate 7. Papillae and the Superficial Epithelium.

A   A light micrograph of the cortical region of the ovary. Small, irregular papillae (P, arrows) are quite evident in cross-section and are observed to contain a core of connective tissue which arises as a villous-like evagination of the tunica albuginea (Ta). Within the subjacent connective tissue of the tunica albuginea, a follicle (F) is also evident. (x 320; estrous rabbit).

B   A cross-section of a papilla as observed by TEM. The cells covering the papilla are polyhedral, contain microvilli, and are interconnected by means of junctional complexes (arrows). A dense core of connective tissue (Ct) composes the axis of this evagination. (x 2,700; estrous rabbit).

C   Some of the cells of the superficial epithelium are irregular in shape, possess few microvilli (Mv), and contain a nucleus which is highly infolded (N). Typically, the mitochondria (M) are small and dense, and the Golgi complex (G) is oriented toward the free surface of the cell. (x 5,500; estrous rabbit).

D   The most commonly observed morphology of superficial cells is illustrated in this transmission electron micrograph. Generally, these cells are rounded and interconnected by means of junctional complexes (Jc). The apices of the cells contain numerous microvilli (Mv), and occasionally, a cilium is observed (as is shown in a tangential section, large arrow). Another characteristic feature is the presence of fluid in intercellular spaces or lacunae (*). (x 12,000; estrous rabbit).

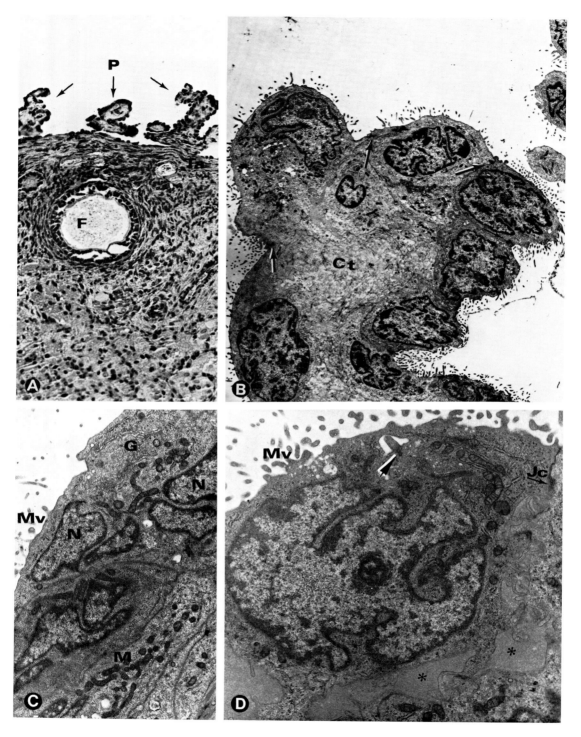

Plate 7. Papillae and the Superficial Epithelium.

Plate 8. A Primordial Follicle.

The primordial (unilaminar) follicle represented in this transmission electron micrograph consists of an oocyte (Oo) surrounded by a single layer of closely apposed, flattened, follicle (granulosa) cells (Fc). At this stage, an assemblage of Golgi membranes (G) and small mitochondria (M) is still located in a juxtanuclear position and likely represent the remnants of the "Balbiani body." Other comparatively small mitochondria, as well as numerous ribosomes and polysomes, are distributed throughout the cytoplasm. The most prominent feature of the oocyte at this stage is a large nucleus (N). In addition, a dense, granular/filamentous material (nucleolar-like bodies or nuage (n)) is evident in some regions of the cytoplasm. The oocyte plasma membrane interdigitates extensively with the membranes of surrounding granulosa cells (arrows), and in some areas, junctional complexes are evident (*arrows). A basal lamina (Bl) surrounds the primordial follicle and delineates it from the adjacent stroma (St) of connective tissue located in the ovarian cortex. (x 4,100; estrous rabbit).

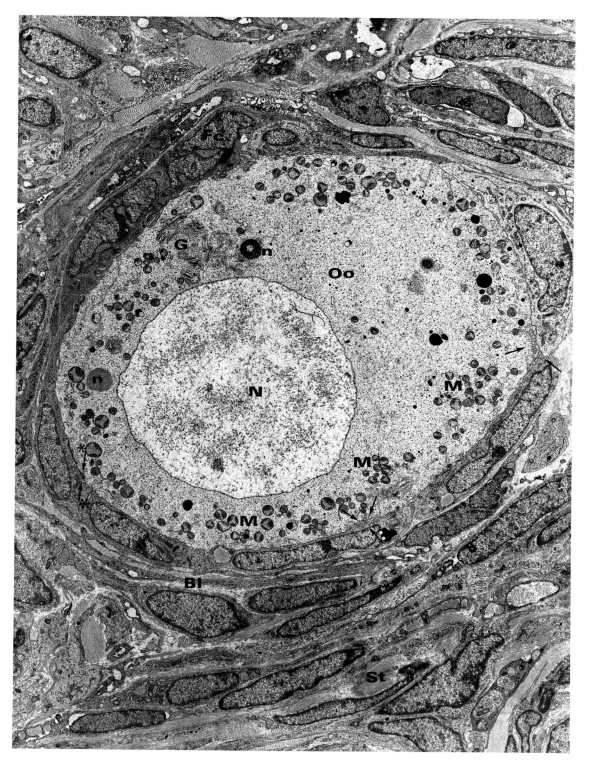

Plate 8. A Primordial Follicle.

Plate 9. Secondary and Early Antral Follicles.

A   A secondary (multilayered) follicle is shown in this light micrograph. Note the many processes which arise from coronal cells and radiate through the zona pellucida (Zp, arrows). The theca interna (Ti) is arranged into a series of flattened, concentric layers characteristic of follicles at this stage of development. N = nucleus; cytoplasm with numerous and scattered organelles = C; follicle cells distributed around the oocyte = Fc; basal lamina = Bl. (1-micron-thick section stained with toluidine blue; x 650; estrous rabbit).

B   An early antral follicle is shown in this light micrograph. The theca interna (Ti) contains numerous polyhedral, epithelioid cells which are secretory in nature (thecal gland) (arrows). The theca externa (Te) has the appearance of connective tissue. A blood vessel (V) and small groups of glandular interstitial cells (Ic) are evident in the stroma (St) which surrounds the follicle. Oocyte = Oo; zona pellucida = Zp; granulosa layer = Gc; antrum = A; adjoining, small primordial follicle = Fp. (1-micron-thick section stained with toluidine blue; x 320; estrous rabbit).

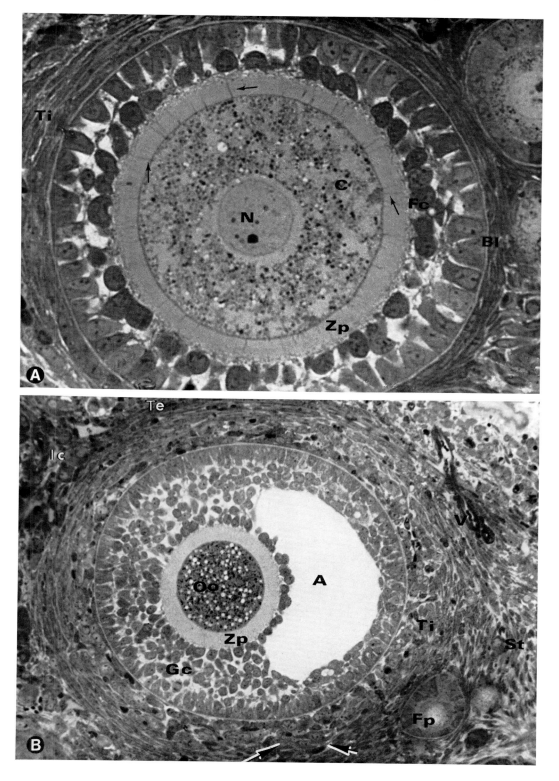

Plate 9. Secondary and Early Antral Follicles.

Plate 10. Call-Exner Bodies.

A   The characteristic appearance of a Call-Exner body (*) present among the granulosa cells of a Graafian follicle is illustrated in this scanning electron micrograph. Follicle cells (Fc) are radially arranged around a spherical cavity which is in turn limited by a thin basal lamina-like structure (Bl). The cavity is filled with liquor folliculi. Antrum of the follicle = A; theca interna surrounding the follicle = Ti.

B   Only a portion of a large Call-Exner body is shown in this transmission electron micrograph. The Call-Exner body illustrated here is quite similar to the one shown in figure A. Follicle cells (Fc) possess long, ameboid-like evaginations which adhere to a thin basal lamina (Bl, arrows). The content of the cavity is composed of follicular fluid (Lf) which is similar in nature to the fluids present in adjacent intercellular spaces and within the antrum of large follicles. Furthermore, a material having a filamentous/reticular texture (*) is also evident within the cavity of the Call-Exner body. (x 4,800 estrous rabbit).

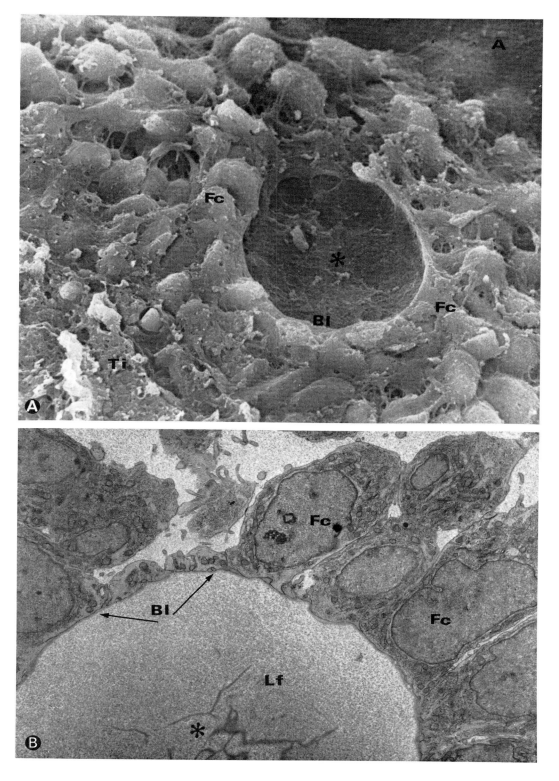

Plate 10. Call-Exner Bodies.

Plate 11. Follicle (Granulosa) Cells.

A As is evident in this scanning electron micrograph, follicle cells (Fc) contained within antral follicles are polyhedral. A basal lamina (Bl) separates the granulosa layer (Gc) from the surrounding theca interna (Ti). (x 2,000; rat in metaestrous).

B The characteristic fine structural appearance of a follicle cell from a multilayered follicle is illustrated in this transmission electron micrograph. The cytoplasm is filled with numerous free ribosomes as well as mitochondria and cisternae of the rough-surfaced endoplasmic reticulum (RER). A single cilium (arrows) is evident in relation to a centriole and a Golgi complex (G). Nucleus with a large, reticulated nucleolus = N and nu, respectively. (x 8,800; estrous mouse).

C A cellular process of a follicle cell which has invaginated into the cytoplasm of an adjacent cell is shown in this micrograph. This particular process contains some organelles and the apposed membranes are interconnected by means of gap junctions (arrows). Such interaction between granulosa cells is often observed in large, multilayered follicles or in preovulatory follicles. (x 6,000; estrous rabbit).

Plate 11. Follicle (Granulosa) Cells.

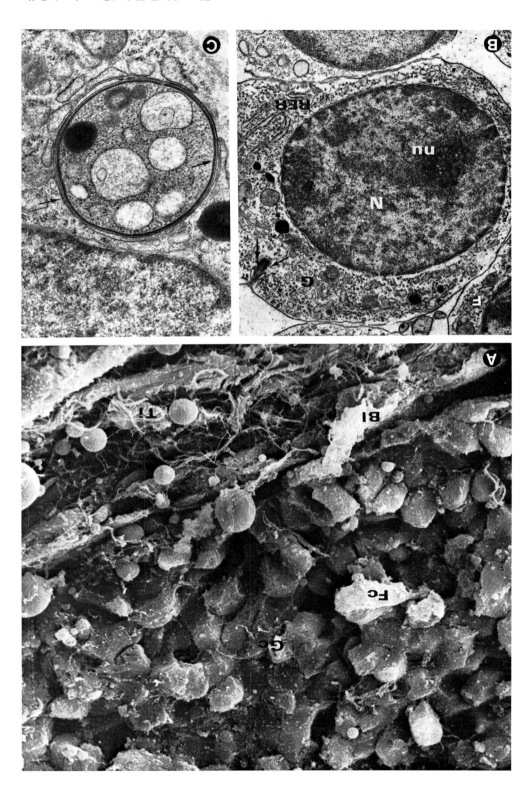

Plate 12. Developing Follicle.

This transmission electron micrograph presents only a partial view of a developing follicle. Within the cytoplasm of the oocyte (Oo), Golgi membranes and mitochondria are primarily located in cortical regions. The nucleus (N) is quite large and contains nucleoli (nu, arrows) which display a reticulated form characteristic of nucleoli active in the synthesis of ribosomal RNA. The zona pellucida (Zp) is present between the oocyte surface and the surrounding follicle cells (Fc). Numerous follicular projections are evident within the zona.

At this stage of development, follicle cells are polyhedral or columnar, have irregularly infolded nuclei, and contain numerous organelles. Typically, cisternae of the rough-surfaced endoplasmic reticulum, mitochondria, and membranes of the Golgi complex are localized in the apical zone of the follicle cell cytoplasm which faces the zona pellucida (see also Plate 14 C.). A basal lamina (Bl) surrounds the follicle, and the ovarian stroma around the basal lamina has begun to become arranged into concentric layers of cells which eventually form the theca interna (Ti). However, at this stage, stromal cells are still spindle-shaped and have not yet differentiated into steroid secreting elements. (x 3,200; estrous rabbit).

Plate 12. Developing Follicle.

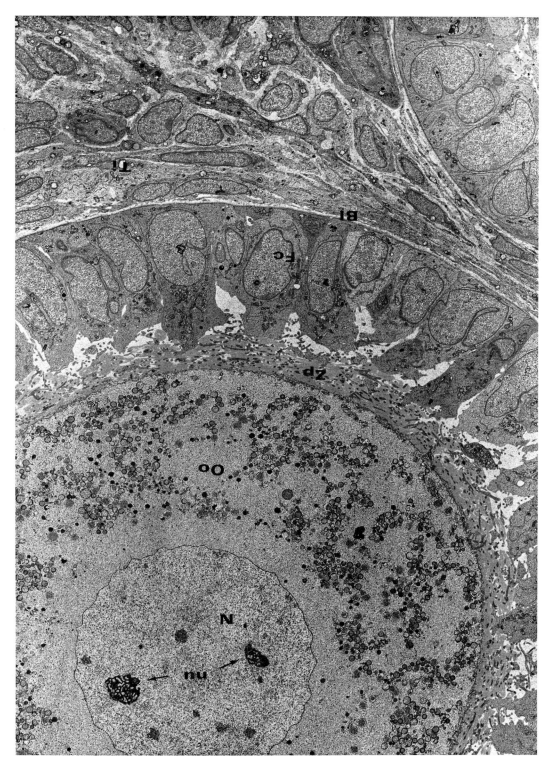

Plate 13. Cumulus Oophorus, Corona Radiata, and Zona Pellucida.

The relationship between the cells of the cumulus oophorus and the zona pellucida is illustrated in this scanning electron micrograph. The surface of the oocyte is not visible as it is completely covered by a thick, dense material – the zona pellucida (Zp). In the particular region presented, some follicle cells (Fc) are partially or completely detached from the zona, thus revealing the distorted, irregular nature of the zona surface. Follicle cells of the corona radiata and the cumulus oophorus contain large, irregular, ameboid evaginations which penetrate obliquely through the zona matrix (arrows). Some of these processes ultimately make contact with the oocyte surface. (x 3,900; estrous rabbit).

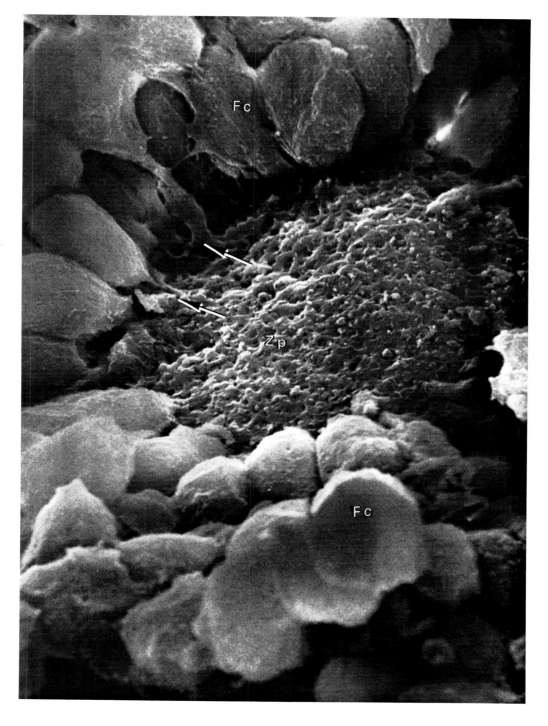

Plate 13. Cumulus Oophorus, Corona Radiata, and Zona Pellucida.

Plate 14. Oocyte and Follicle Cells.

A   Surface section of the cumulus oophorus of a secondary follicle as seen by SEM. Nucleus of oocyte = N; cytoplasm of oocyte containing numerous vacuoles, some of which correspond to mitochondria, while others are likely Golgi membranes = C; follicle cells = Fc; zona pellucida = Zp. (x 1,250; estrous rat).

B   High-magnification scanning electron micrograph of the exposed surface of the zona pellucida in a large Graafian follicle. Numerous crypts and invaginations (arrows) are evident on the surface of the zona, and the zona material appears to be composed of a dense, amorphous material (*). A fine coating of granules is also present on the surfaces of the zona (arrows plus*). (x 6,000; estrous mouse).

C   The relationship between follicle cells and the zona pellucida (Zp) is illustrated in this transmission electron micrograph. The cytoplasm of coronal cells (Fc) contains a dense population of ribosomes, polysomes, and, in addition, numerous mitochondria, Golgi vesicles, and elongated and extended elements of the rough-surfaced endoplasmic reticulum. Ameboid processes (arrows), packed with microfilaments (see Plate 15 B), extend from the follicle cells to the surface of the oocyte. (x 4,700; estrous mouse).

44

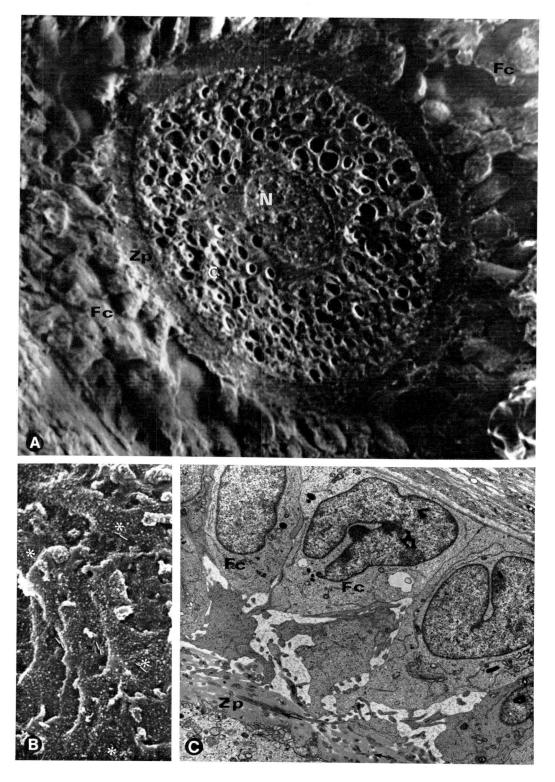

Plate 14. Oocyte and Follicle Cells.

Plate 15. Zona Pellucida and Coronal Cells in a Mature Follicle.

A   High-magnification stereo view of the zona pellucida (Zp) and surrounding coronal cells in a mature mouse follicle. The coronal cells (Fc) possess long and irregular ameboid evaginations which penetrate obliquely into the zona (arrow). The follicle cells are covered with a material which has a granular/filamentous texture. Numerous crypts and invaginations are evident on the zona surface. (x 6,760; estrous mouse).

B   This transmission electron micrograph demonstrates the typical contact between the plasma membrane of the oocyte (Oo) and a follicle cell process. In the process, both bundles of microfilaments (Mf) and clusters of granules (arrow 1) are quite evident. Granules of the same morphological appearance are present on the surface of the process (arrow 2) and within the zona matrix (arrow 3). Zona pellucida = Zp.(x 80,000; rabbit at 4 hours postcoitum).

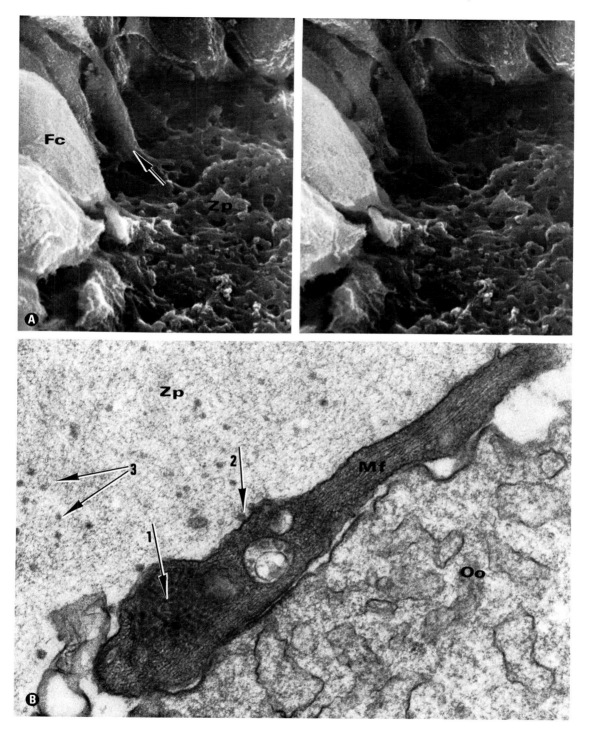

Plate 15. Zona Pellucida and Coronal Cells in a Mature Follicle.

Plate 16. Theca Interna and Interstitial Cells.

A   A scanning electron micrograph of a vesicular follicle illustrating the interior surface of the stratum granulosum and the flattened cells which cover it (Gc). The follicle is surrounded by the theca interna (Ti, arrows) and lies in close proximity to cortical areas of the ovary which are covered by the superficial epithelium (Se). (x 200; estrous mouse).

B   This transmission electron micrograph demonstrates the differentiated cells of the theca interna gland, which is present as a series of layers around large, developing follicles (F). The cells of the theca interna (Ti) have an epithelioid appearance and are in close proximity to a capillary (Ca). This specimen was stained with ruthenium red. (x 3,200; mouse in diestrous).

C   A relatively high-magnification view of a typical, differentiated, theca interna gland cell. The cytoplasm of these cells contains numerous lipid droplets (L) and mitochondria (M) which possess tubular and/or villiform cristae. The plasma membrane is thickened due to the presence of a continuous cell coat or glycocalyx (arrows). An erythrocyte (E) is evident in the lumen of a capillary. Nucleus = N. This specimen was stained with ruthenium red. (x 8,500; mouse in diestrous).

D   High magnification of the cytoplasm of an interstitial cell. Lipid droplets (L) and mitochondria (M) with villiform cristae are typical aspects of this type of steroid-secreting cell. Lysosome-like body = Ly. Compare the features of this cell with those of a theca interna cell – in general, they are quite similar. (x 27,000; mouse in estrous).

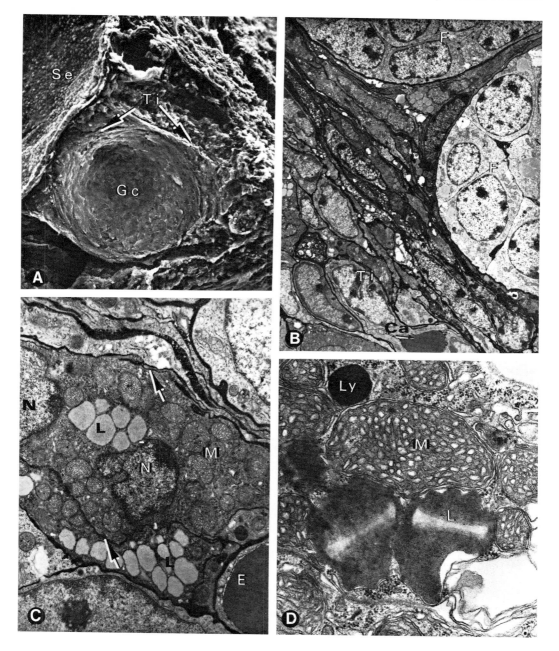

Plate 16. Theca Interna and Interstitial Cells.

Plate 17. Ovarian Stroma and Smooth Muscle Cells.

A   In the cortical areas of the ovary, connective cells (fibroblasts and fibrocytes) (Fi) intermingle with bundles of collagen fibers (Cf) forming a dense stroma. (x 1,820; estrous mouse).

B   This transmission electron micrograph illustrates the tunica albuginea of the ovary. Several fibrocytes (Fi) and bundles of collagen fibers (Cf) are evident in this section. Superficial epithelium = Se. (x 7,500; proestrous rat).

C   Within the ovarian stroma and intermingled among some fibrocytes (Fi), a rather ramified process possibly of a smooth muscle cell (myoid cell), is evident (*). Some mitochondria and numerous bundles of microfilaments are present within the cytoplasm of this contractile cell. The insert illustrates an area of close contact between a nerve terminal and the cytoplasm of the smooth muscle cell (arrows). (x 13,500; *insert*, 20,000; estrous mouse).

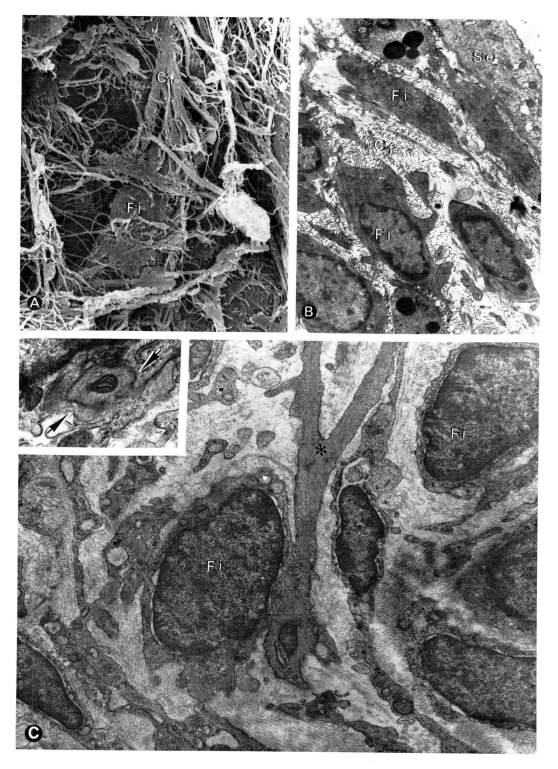

Plate 17. Ovarian Stroma and Smooth Muscle Cells.

2   The Ovary and Ovulation

2.1. The Follicular Stage

Generally, it is assumed that when an ovarian follicle reaches the Graafian stage of development, it is prepared for ovulation. As previously mentioned (Chapter 1), the initial event associated with the development of a follicle to the preovulatory stage is the rapid increase in the secretion and accumulation of liquor folliculi. Accompanying this secretion is not only a marked swelling of the antrum, but also a series of successive morphological changes (1) in the oocyte, (2) in each of the cellular layers of the follicle, and especially (3) in the portion of the follicle related to the ovarian surface – the apex of the follicle. In many mammals, these changes take place approximately 10 to 15 hours prior to the rupture of the follicle. In mammals that are reflex ovulators, such as the cat and rabbit, morphological transformations within the follicle are readily studied under normal physiological conditions, since ovulation occurs within a few hours following coitus. However, in the majority of mammals, the analysis of ovulation under *normal* conditions requires the ability to time the estrous cycle and then to examine patiently a fairly sizable number of ovaries.

Although we understand that for most mammals the induction of ovulation by exogenous gonadotropins greatly facilitates experimental investigation, since both the "usual" number as well as an increased number (superovulation) of ovulation sites may be obtained, we would like to emphasize that when subtle morphological processes are being studied, exogenously induced versus natural ovulation may not necessarily be strictly comparable. Consequently, the observations presented in this chapter arise from the examination of animals undergoing natural ovulation, and it is from these animals that our interpretations and speculations are derived.

2.1.1. The Preovulatory Follicle

The conspicuous growth of the follicle, the swelling of the antrum, and the thinning of the granulosa layer that occur during the preovulatory phase are primarily the result of a rapid increase in follicular fluid (Plate 18 A). At this time, the liquor folliculi is markedly less viscous than previously and is composed of both the secretions of granulosa cells and plasma proteins which enter the follicle by transudation *(Burr and Davies,* 1951; *Caravaglios* and *Cilotti,* 1957; *Motta,* 1965). The leakage of plasma proteins into the antrum is likely a consequence of both the increased vascularization of the preovulatory follicle and increased permeability of thecal capillaries *(Burr and Davies,* 1951; *Christiansen, Jensen,* and *Zachariae,* 1958).

As a consequence of the rapid accumulation of fluid within the antrum, numerous granulosa cells (originating primarily from the stratum granulosum but also from the cumulus oophorus) become dissociated from each other, lose their intercellular contacts, or remain connected by thin strands of cytoplasm in which small junctional complexes are still evident (Plate 18 A for general appearance) *(Motta, Takeva,* and *Nesci,* 1971; *Mestwerdt, Müller,* and *Brandau,* 1977). Further contact between granulosa cells is also maintained by means of processes that invaginate into the cytoplasm of adjacent cells, forming gap junctions at the interface of the two plasma membranes *(Espey* and *Stutts,* 1972; *Merk, Albright,* and *Botticelli,* 1973; *Albertini* and *Anderson,* 1974; *Albertini, Fawcett,* and *Olds,* 1975; *Mestwerdt, Muller,* and *Brandau,* 1977). The dissociation of granulosa cells is most pronounced in the region of the cumulus oophorus nearest the apex of the follicle. The intercellular spaces within the theca interna also become distended by fluids that likely originate from surrounding capillaries *(Bjersing* and *Cajander,* 1974 a; *Parr,* 1974; *Cherney, DiDio,* and *Motta,* 1975). (Plate 18 B). The dissociation of the granulosa layer in the preovulatory follicle seems to be caused by components of plasma that enter the granulosa through large gaps in the basal lamina and become mixed with liquor folliculi initially secreted by granulosa cells *(Motta, Cherney,* and *DiDio,* 1971; *Parr,* 1974; *Cherney, DiDio,* and *Motta,* 1975).

The infiltration of fluids into the preovulatory follicle is usually restricted to the apical portion of the follicle. Just prior to ovulation, the appearance of these fluids seems not only to be correlated with the dissociation of granulosa cells, but also to be correlated with the distention and disorganization of other cellular layers such as the theca, tunica albuginea, and the superficial epithelium (Plates 19 A, 19 B, 20). The disorganization of these structures likely contributes to the characteristic distended appearance of a preovulatory follicle (Plates 21, 22). Furthermore, in follicles estimated to be in the final stages of preovulatory development, blood cells (erythrocytes, leukocytes, and platelets) are observed to be closely packed within the lumen of dilated capillaries (Plate 18 B) (*Byskov*, 1969; *Motta, Cherney,* and *DiDio*, 1971; *Bjersing* and *Cajander*, 1974 a, 1974 b; *Parr*, 1974; *Cherney, DiDio,* and *Motta*, 1975). Immediately prior to ovulation, blood cells are also observed among the dissociated tissues of the follicle and, occasionally, within the antrum where they mix with free granulosa cells (*Motta, Cherney,* and *DiDio*, 1971; *Bjersing* and *Cajander*, 1974 a, 1974 b; *Parr*, 1974; *Cherney, DiDio,* and *Motta*, 1975).

Most of the granulosa cells in the area corresponding to the anticipated site of follicular rupture are flattened, have lost their intercellular contacts, and have become detached from the granulosa layer. By contrast, granulosa cells in deeper regions of the preovulatory follicle retain their characteristic polyhedral shape as well as their numerous intercellular contacts. The granulosa wall in these deeper regions may become slightly plicated, initiating a process that is not fully completed until the follicle is transformed into a corpus luteum (Plate 18 A, and section 2.3).

Within a few hours or even moments before ovulation, the cumulus oophorus is usually detached from the stratum granulosum and may be observed floating freely in the antrum. The detachment of the oocyte from the granulosa layer likely involves the mechanical dissociation of cells as a result of the passive infiltration of fluid into the antrum. In addition, the ameboid contractions of granulosa cells, which contain extensive microfilamentous networks composed of the contractile proteins actin and myosin, could contribute to the detachment of the cumulus oophorus (Chapter 1, Plate 15; Plates 19 C, 19 D) (*Motta* and *DiDio*, 1974; *Motta* and *Van Blerkom*, 1975; *Cavallotti, DiDio, Familiari, Fumagalli,* and *Motta*, 1975). Generally, just prior to ovulation, the processes of coronal cells that traverse the zona pellucida and contact the surface of the oocyte are withdrawn (see Chapter 1, Plate 15, and Plate 18 C) (*Sotelo* and *Porter*, 1959; *Odor*, 1960; *Motta* and *Van*

Blerkom, 1974). It is not unlikely that the mechanism of retraction involves the action of microfilaments present in the granulosa cell processes.

As observed by scanning electron microscopy, and without significant variation among the numerous mammals studied, preovulatory follicles are blister-like structures that protrude markedly from the ovarian surface (Plates 21, 22, 23) (*Motta, Cherney,* and *DiDio*, 1971; *Nilsson* and *Munshi*, 1973; *Bjersing* and *Cajander*, 1974 a, 1974 b; *Motta* and *Van Blerkom*, 1975). At the base of a preovulatory apex, the superficial epithelium is composed of cuboidal or polyhedral cells (Plates 21 B, 22 A) containing a large population of microvilli and, frequently, an isolated cilium (Plate 24 A). By contrast, and indeed as may be expected, the lateral surfaces of the protruding follicle consist of a superficial epithelium whose cells, straightened by tension, appear rather elongated and with a reduced number of microvilli (Plates 21 B, 22 A, 24 B). The apices of preovulatory follicles possess large areas of squamous cells containing scarce, short microvilli. In regions adjacent to the apex, numerous superficial cells lose contact with one another, undergo pronounced degenerative alterations, and begin to "slough off" from the surface of the follicle (Plates 22 B, 23 A, 24 C) (*Blandau*, 1967; *Motta, Cherney,* and *DiDio*, 1971; *Bjersing* and *Cajander*, 1974 a, 1974 b; *Motta* and *Van Blerkom*, 1975). The desquamation of the superficial epithelium at the apex of a preovulatory follicle appears to be the result of necrotic or cytolytic processes taking place in subjacent cortical areas (see section 2.2), and as observed by scanning electron microscopy, desquamation of the apical superficial epithelium exposes the underlying connective tissue (Plates 23 A, 24 D, 25 A).

At the apices of some preovulatory follicles, irregular, fluid-like drops are seen either free on the surface of cells or in the process of being extruded from the cytoplasm and/or intercellular spaces among the superficial cells (Plate 24 B) (*Cajander*, 1976). When these same areas are studied in thin sections by transmission electron microscopy, a large accumulation of an amorphous, fluid-like material is noted in the distended intercellular spaces between cells of the superficial epithelium (Plate 20 A), as well as beneath the basal lamina and in the connective stroma of the tunica albuginea (Plate 20) (*Motta* and *Van Blerkom*, 1975). In areas adjacent to the apex, cells composing the tunica albuginea and thecal layers often show clear signs of vacuolization and degeneration. In regions where fluid accumulation is particularly abundant, these degenerating elements are almost "floating" in this material, as are blood cells and cellular debris. This fluid-like material, which permeates the intercellular spaces of the theca, tunica albuginea, and superficial epithelium, seems to be correlated with the disruption of collagen fibrils, which leads to a marked weakening of the apical wall. In addition, a reduction in blood flow to a circumscribed region on the external surface of the follicle probably contributes to lysis of the thecal and granulosa cells in that area (*Blandau*, 1970). Finally, just prior to ovulation, antral fluid is observed to mix with this amorphous material as a direct consequence of the disorganization of the basal lamina which separates the granulosa and thecal layers (Plates 19A, 19B, 20).

2.1.2. The Maturation of the Oocyte

Coincident with the final development of a preovulatory follicle, the oocyte, which until this point has remained in the prophase of meiosis (dictyate stage), resumes meiosis approximately 9 to 12 hours before ovulation (this figure is variable and species-dependent). The culmination of the first meiotic or maturative division is the discharge of the first polar body, which may remain attached to the oocyte for a brief period of time by a cytoplasmic bridge (*Zamboni*, 1971). The first meiotic metaphase lasts approximately 6 hours in the mouse, rat, hamster, and rabbit and about 20 hours in the pig. Usually, the first meiotic division occurs before ovulation, while the second maturative division is not completed until or unless fertilization has taken place. Notable exceptions to this pattern of oocyte maturation include the dog and fox, in which the first polar body is not extruded until several days after ovulation, and the insectivorous mammals, in which spermatozoa enter the ovarian follicle and penetrate oocytes that are then ovulated in the pronuclear stage of development (*Hafez*, 1970). The results of the two maturative divisions, when and if completed, are the production of polar bodies and an ovum with a haploid complement of chromosomes.

At both the molecular and cellular levels, major changes occur within the oocyte during the maturative stages immediately preceding ovulation. Significant qualitative and quantitative alterations in the pattern of protein synthesis have been reported for maturing mouse (*Schultz* and *Wassarman*, 1977), rabbit (*Van Blerkom*, 1977; *Van Blerkom* and *McGaughey*, 1978), pig (*McGaughey* and *Van Bler-*

kom, 1977), and sheep oocytes (*Warnes, Moor,* and *Johnson*, 1977). In addition, RNA synthesis continues up to at least germinal vesicle (nuclear) breakdown (*Wassarman* and *Letourneau*, 1976). Taken together, these studies demonstrate that the mammalian oocyte is quite active in macromolecular syntheses during the relatively brief period of maturation.

At the cellular level, the formation of electron-dense granules from elements of the Golgi complex and the distribution of these granules within the cortical region of the oocyte cytoplasm (cortical granules) are major features of the preovulatory phase. The meiotic stages leading to telophase and the separation of the first polar body are characterized by the fragmentation and disappearance of the nuclear membrane (germinal vesicle) and nucleoli, as well as by the liberation of chromosomes into the cytoplasm (*Zamboni*, 1971). Light and transmission electron microscopic studies have demonstrated that in contrast to most animal cells, the oocytes of many species undergo cytokinesis in the absence of centrioles or astral rays (*Odor* and *Blandau*, 1951; *Zamboni* and *Mastroianni*, 1966; *Hertig* and *Adams*, 1967). In place of centrioles, clusters of compactly arranged vesicles are located at the poles of the metaphase spindles (*Zamboni*, 1970, 1971).

2.2. Ovulation

At the moment of ovulation, large, irregular areas of the apex of a follicle are ruptured. It is rather difficult to view the oocyte and corona radiata in the process of ovulation directly by scanning electron microscopy, since an ovulating or recently ovulated follicle is usually covered by a large quantity of a viscous material and cellular debris – the follicular fluid expelled with the oocyte (*Motta* and *Van Blerkom*, 1975; *Cajander*, 1976; *Rosenbauer, Jansen,* and *Lindauer*, 1976). In only a few cases, and after a laborious examination of more than 200 different ovaries in which ovulation was occurring under natural conditions (in estrous mice, rats, and in rabbits following coitus), was it possible to obtain satisfactory images of the oocyte in this process. Clearly, a full appreciation of the dynamic sequence of events taking place during ovulation cannot be captured in the static images of the electron microscope, but rather requires other techniques such as microcinematography. However, the micrographs presented in Plates 25, 26, and 27 likely represent critical sequences of this dynamic process, even taking into consideration the probable presence of some minor artifacts. These artifacts, such as the detachment of some coronal cells and partial denudation of the zona pellucida (Plates 25 A, 26), are fortunate since they permit the direct observation of the oocyte, which normally would be obscured from view by the mass of cumulus cells. As may be observed in Plates 25 A and 26, large quantities of a fluid-like material and cellular debris are present around the ruptured apex of the follicle, and the oocyte appears to emerge from within the cavity of the follicle in a mass of liquor folliculi. The fluids surrounding the oocyte seem to be in a condensed, possibly coagulated form and to contain clusters of granulosa cells (probably also those cells which have become detached from the zona pellucida) which collectively contribute to a partial obscuring of the subtle architecture of the oocyte.

2.2.1. The Process of Ovulation

Mammalian ovulation would outwardly appear to be a rather simple phenomenon if regarded merely as a dynamic sequence of events. However, when it is considered that ovulation involves a series of timed hormonal, biochemical, biophysical, and morphological events, the phenomenon becomes more complicated. Although there are no doubts about the hormonal requirements for ovulation, the concomitant biophysical and morphological processes are not well understood. During the past 110 years, numerous and sometimes truly ingenious theories have been advanced in an effort to explain ovulation in terms of a unified concept (see reviews by *Hisaw*, 1947; *Watzka*, 1957; *Asdell*, 1962; *Blandau*, 1967; *Rondell*, 1970; *Espey*, 1974; *Weir*, 1977; *Edwards et al.*, 1977). The following list represents some of the theories proposed over almost a century of research in an attempt to define what "causes" a follicle to ovulate:

1. an increase in intrafollicular pressure prior to ovulation (a theory currently rejected since it is now clear that follicular pressure does not increase in the preovulatory follicle (*Rondell*, 1970; *Espey*, 1974),
2. the contraction or "squeezing" of the follicle by smooth muscle cells,
3. the enzymatic digestion of the follicular apex,
4. vascular changes within the follicle, and
5. nervous control of preovulatory events.

At present, no single theory is widely accepted as being the "actual" cause of ovulation, although

neural control of enzymatic and contractile activity has received some substantial experimental support (see *Rondell*, 1970; *Espey*, 1974).

From a strictly morphological point of view, it has become increasingly apparent that just prior to the dehiscence of the follicle, a progressive degenerative alteration in the cellular layers composing the apex of the follicle takes place (*Espey*, 1967; *Byskov*, 1969; *Motta*, *Cherney*, and *DiDio*, 1971). As observed primarily by transmission electron microscopy, this alteration consists of a gradual decomposition of the intercellular ground substance of the connective tissue with a concurrent dissociation and degeneration of the fibrillar and cellular components. These components include superficial cells, the tunica albuginea, and thecal and granulosa cells in the immediate vicinity of the follicular apex. Furthermore, in follicles estimated to be in the immediate preovulatory stage, extraluminal red blood cells and platelets, originating from the rupture of small vessels, are present in the cellular structures of the apex (*Motta*, *Cherney*, and *DiDio*, 1971; *Bjersing* and *Cajander*, 1974a, 1974b; *Parr*, 1974). The dissociation of fibers and cells, as well as the labilization of the ground substance, appear to be enhanced by fluid infiltrating into perifollicular zones (*Motta*, *Cherney*, and *DiDio*, 1971) and accumulating primarily under the elongated cells of the apical superficial epithelium (*Cherney*, *DiDio*, and *Motta*, 1975; *Parr*, 1974).

Observations of preovulatory follicles by scanning electron microscopy demonstrate that cells of the apical superficial epithelium lose their microvilli, become progressively flattened, and ultimately slough off from the apex (*Motta*, *Cherney*, and *DiDio*, 1971; *Bjersing* and *Cajander*, 1974a, 1974b). A thin "fluid" layer covers the stigma of mouse preovulatory follicles (*Nilsson* and *Munshi*, 1973), whereas in the rat and rabbit, as well as in the mouse, scattered drops of a fluid-like material are observed on the surface of the apex (*Motta* and *Van Blerkom*, 1975; *Cajander*, 1976). These drops apparently arise by seepage from underlying intercellular pools (Plates 20, 24 B) and reach the follicular surface through degenerating regions of the superficial epithelium. From a reconstruction of the submicroscopic, morphological events that occur just prior to ovulation, it seems evident that a pronounced increase in fluids (edema) in the cellular components of the apical wall may be the major and final observable "cause" of the rupture of the follicle. The apical wall has already been significantly thinned and weakened, most likely as a result of enzymatic activity (*Espey*, 1974) and/or reduced blood flow (*Blandau,* 1970). The origin of the fluids in the perifollicular area may be related to an increase in the vascularity of the preovulatory follicle (*Burr* and *Davies,* 1951) with a concomitant increase in the permeability of thecal capillaries (*Christiansen, Jensen,* and *Zachariae,* 1958). An increase in the permeability and fragility of the endothelial (capillary) wall may depend upon the local synthesis and release of steroid hormones (estrogens and progesterone) by perifollicular components (primarily the theca interna and partially luteinized granulosa cells) and would lead to local edema and hemostasis (*Rona,* 1963; *Szego* and *Gitin,* 1964). Thus, as a direct result of this edema, the apex of the follicle, which is already under tension owing to the increase in the fluid volume of the antrum, becomes so distended and fragile that the dehiscence of the follicle is the only possible consequence.

The role of lysosomes in ovulation has also received attention. Recently, *Cajander* and *Bjersing* (1975, 1976) have demonstrated the lysosomal nature of many of the dense bodies that populate the cytoplasm of the surface epithelium prior to ovulation in the rabbit. On the basis of these observations it was suggested that the superficial epithelium of the ovary may have an important role in liberating "proteolytic ovulatory enzymes" that may be responsible for the degeneration of the apex and consequent ovulation. Furthermore, these investigators speculated that the labilization caused by lysosomes could be induced by prostaglandins, whose presence has been implicated in the ovulatory process (*Yang, Marsh,* and *LeMaire,* 1974). However, *Rawson* and *Espey* (1977) suggest that the accumulation of dense bodies in the surface epithelium of the rabbit ovary is not essential to ovulation, but rather may reflect a physiological response to traumatic changes which occur in a mature follicle as it approaches rupture.

Keeping in mind the morphological events described both in Chapter 1 and in the above discussion, we should like to present some theoretical considerations concerning ovulation. Ovulation is one step in a continuum of developmental and differentiative processes initiated with the growth of a primordial follicle and progressing through the transformation of an ovulated follicle into a corpus luteum; however, if fertilization has not taken place, it ends with the regression of the corpus luteum and formation of a corpus albicans. The initial and terminal stages are somewhat arbitrarily determined as a matter of convenience and for descriptive purposes. The entire continuum is what we term the luteo-follicular complex. As an example of this continuum, even during the *preovulatory* stages, some granulosa cells undergo precocious luteinization (and are able to secrete steroid hormones), thus heralding the postovulatory role of the follicle as a glandular structure (see section 2.3). Likewise, the highly vascular theca in the preovulatory follicle begins to exert pressure on the overlying granulosa layer, initiating a folding or plication of the granulosa which is not completed until after ovulation. These and other morphophysiological processes continue in a follicle even if ovulation has not taken place. Thus, it appears probable that ovulation occurs only if a follicle is in the process of transforming into a corpus luteum during the preovulatory stage, and then only if such a follicle is in a superficial position in the ovarian cortex with enough of the weakened follicular wall exposed to permit rupture. If such a follicle is located in a more interior region of the stroma, the oocyte may not be expelled, and while the follicle develops into a corpus luteum (termed false corpus luteum or atretic follicle; see Chapter 1), the oocyte rapidly degenerates.

Therefore, rather than an "explosive" event, ovulation seems to be a gradual process in which the oocyte, immersed in a gel-like material, escapes from the fractured surface of the ovary. This gel-like material is composed of antral fluids (liquor folliculi), fluids that had accumulated in intercellular spaces, granulosa cells, and cellular debris. A dynamic reconstruction of morphological observations obtained by transmission and scanning electron microscopy agrees with the microcinematographic demonstration that ovulation *in vivo* is a relatively slow process (*Blandau,* 1967).

If ovulation simply consists of the dehiscence of the apical portion of the follicle, then one question central to an understanding of this entire process is what forces are involved in the expulsion of the oocyte and the follicular fluids? On the basis of numerous morphological and physiological observations, the following contributory mechanism may be proposed. Just prior to the rupture of the follicle, the cumulus oophorus is generally detached from the thinned and extensively dissociated stratum granulosum. At this point, the oocyte lies free in the liquor folliculi, surrounded by an irregular mass of cells that are radially arranged around the oocyte (the corona radiata) (*Gwatkin* and *Carter,* 1975). These granulosa cells display numerous elongated, ameboid evaginations (*Chang, Anderson, Lewis, Ryan,* and *Kang,* 1977). The contractile or pulsating movements of the cells (*Motta* and *DiDio,* 1974), even if limited in nature, could be a contributing factor in the migration of the cumulus mass to the stigma of the follicle and, from the stigma to the overlying, adjacent fimbrae of the oviduct. The presence of smooth muscle cells, especially in the perifollicular regions, may well be the source of the "squeezing of the follicle" which occurs during ovulation (for references on smooth muscle cells, see *Fumagalli* and *Motta,* 1969; *O'Shea,* 1970; *Osvaldo-Decima,* 1970; *Amsterdam, Lindner,* and *Groschel-Stewart,* 1977; and Chapter 1). Furthermore, the peristaltic activity of the oviduct (*Talo,* 1974) and the ciliary movements on the mucosal surface of the fimbrae (*Odor* and *Blandau,* 1973) may have a significant role in creating local currents that ensure the complete expulsion of the oocyte and the contents of the follicular cavity. Finally, the presence of nerves in the perifollicular region, and the probable existence of functional neuromuscular junctions, suggest that ovulation may be at least partially under neural control (*Walles, Edvinsson, Owman, Sjoberg,* and *Sporrong,* 1976).

2.3. The Luteal Stage

2.3.1. The Corpus Luteum – General Considerations

As mentioned in the previous section, the transformation of a follicle into a corpus luteum usually begins before ovulation (generally by several hours), and is initiated under the influence of pituitary gonadotropins (particularly luteinizing hormone), which are relatively abundant in the circulating blood at this time. Evidence of precocious luteinization of granulosa cells has been observed by transmission electron microscopy in several species. Fine structural changes associated with luteinization include the appearance of smooth-surfaced endoplasmic reticulum, lipid droplets, and mitochondria with villiform cristae (for the rabbit, see *Blanchette,* 1966a, 1966b; for the mouse and rat, see *Hadjioloff, Bourneva,* and *Motta,* 1973, and *Björkmann,* 1962, respectively). Furthermore, histochemical and cytochemical observations demonstrate the presence of many of the enzymes required for steroid synthesis in the granulosa cells of preovulatory follicles (*Bjersing* and *Carstensen,* 1964; *Blanchette,* 1966a; *Bjersing,* 1967b). After ovulation, when the corpus luteum is fully differentiated, its structure and function are those of a typical endocrine gland with progesterone (and some estrogen), the primary secretory product. It is the secretion of steroid hormones by the corpus luteum that (1)

prepares the endometrium for implantation, (2) enables the blastocyst to implant, and (3) maintains pregnancy after implantation. The functional life span of a corpus luteum is completely dependent upon the fate of the embryo. If implantation has taken place, the corpus luteum undergoes extensive enlargement and continues to synthesize and secrete steroid hormones (the corpus luteum of pregnancy). By contrast, if an embryo fails to implant (or if an oocyte is not fertilized), the corpus luteum undergoes atrophy and eventual degeneration (the corpus luteum of menstruation in humans). In both situations, the developmental changes leading to the formation of the corpus luteum are similar, as are the subsequent regressive events leading to the formation of a small white nodule termed the corpus albicans.

The functional life span of a corpus luteum also varies greatly with the species. For example, a corpus luteum persists for about 28 days in a pregnant rabbit and about 2 to 3 months in a pregnant human. However, if implantation has not taken place, the corpus luteum (or corpora lutea in rabbits) remains functional for only 14 days in both species. Additional stimulation for the synthesis and secretion of progesterone by the corpus luteum of pregnancy comes from the presence of lactogenic hormone. The "life-cycle" of the human corpus luteum, as examined by histological techniques, has been clearly described by *Meyer* (1911) in a classic study which to this day remains central to any examination of the corpus luteum. Other reviews that may be of benefit to the reader and that offer both traditional and contemporary insights into this subject are the works of *Corner* (1919), *Momigliano* (1927), *Pratt* (1935), *Motta* (1936), *Dubreuil* and *Riviere* (1947), *Brambell* (1956), *Corner* (1956), *Watzka* (1957), *Harrison* (1962), *Mossman* and *Duke* (1973), and *Balboni* (1976). The transformation of a ruptured follicle into a corpus luteum involves characteristic foldings of the granulosa layer toward the central portion of the residual cavity. The plications of the granulosa layer possess a highly vascular connective tissue core which arises from the surrounding theca. The plications of the granulosa layer, together with the contraction of smooth muscle cells in the perifollicular region, significantly reduce the size of the follicular cavity after ovulation.

2.3.2. Repair of the Ovulated Follicle

When the apex of the follicle has been ruptured and the cumulus mass and oocyte have been discharged, a small mass of blood (from small vessels that hemorrhaged at ovulation) and follicular fluids is formed (Plates 28 B, 29). The mass forms rather rapidly and acts as a "stopper" to seal the residual cavity after the discharge of the oocyte (Plates 25 B, 27, 28, 29). Within this sealed structure, intact vessels and connective cells from the surrounding theca begin to proliferate and subsequently penetrate the basal lamina. Granulosa cells, which remain in the cavity, become markedly enlarged and polyhedral elements. The granulosa cells, which are completely dissociated from each other, begin to migrate and rearrange themselves into epithelial-like structures composed of anastomosing cellular networks in close association with groups of small blood vessels. Thus, cells from the stratum granulosum and small blood vessels from the theca perforate and ramify through the blood/fluid mass formed after ovulation. The mass is gradually replaced by a highly vascular cellular structure, and as a consequence of these morphogenetic events, the ruptured follicle is transformed into a solid, glandular body – the corpus luteum (Plate 35 A) (a yellowish structure usually visible to the unaided eye).

As observed by scanning electron microscopy, the mass on the surface of a recently ovulated follicle is an amorphous prominence to which are attached debris, granulosa cells (Plates 28 A, 29), and numerous flattened connective cells (Plates 28 A, 29, 30 A). These connective elements are likely fibroblasts and/or other connective cells (macrophages) originating from the adjacent and subjacent connective stroma and which migrate to the apical regions of the ruptured follicle. These cells usually possess an ovoidal surface (corresponding to the region of the cytoplasm containing the nucleus) from which extend numerous thin, cellular projections (Plates 30 B, 31 A). The cellular projections form an extensive network and appear either enmeshed in or associated with a thick layer of fibrils on the surface of the mass (Plate 31 A). In other regions of the mass, the dense packing of fibrils forms a "felt-like" layer in which are immersed scattered fibroblasts (Plate 31 B). These micrographs probably demonstrate the actual deposition of collagen (collagenogenesis) by the fibroblasts. The deposition of collagen in the postovulatory follicle seems to involve the secretion of tropocollagen, the formation of elementary fibrils, and the eventual coalescence of these fibers into bundles which then occupy intercellular spaces. The fibrillar network may represent a substratum upon which subsequent cellular migrations are accomplished (*Van Blerkom* and *Motta*, 1978).

Typically, at approximately two days following ovulation, the apices of many ruptured follicles are covered by a rather continuous layer of connective tissue and intercellular fibers. Concurrently, a layer of extremely flattened cells begins to migrate over the exposed connective tissue, such that subjacent fibroblasts protrude markedly (Plates 30 B, 31 A, 31 C). From a dynamic, morphological reconstruction of numerous scanning electron micrographs, it appears that in the majority of follicles, this layer is composed of elements of the superficial epithelium that proliferate at the basal areas of the follicle and then migrate over the connective tissue matrix formed at the disrupted regions of the follicle after ovulation (Plate 33). At higher magnifications, "zipper-like" interdigitations of the cellular borders of these migrating cells are clearly evident (Plate 31 C). The process of proliferation and migration of the superficial epithelium from basal to apical areas of the ruptured follicle is one of the most remarkable events to occur on the surface of the postovulatory follicle. In the mammal, a general impression of this process may be obtained from the micrographs presented in Plate 33. Elements of the basal superficial epithelium appear to almost "flow" toward the ruptured lateral and apical regions which are already covered by connective tissue. Transition forms of the superficial epithelium, from the microvillus, polyhedral cells to the extremely flattened, relatively microvilli-free elements, are evident in Plate 33 B. The association between the underlying connective tissue and the apical superficial epithelium is shown in Plate 33 C. In addition, apical protrusions arising from the superficial cells located around the base of the follicle are frequently observed (Plate 32). Although the significance of these protrusions is obscure, their appearance is quite similar to apical protrusions observed in the uterus during the preimplantation stages of pregnancy and, at present, thought to be associated with endocytotic processes (see Chapter 3). In any event, in the rabbit, they are likely a manifestation of intense cellular activity associated with the repair of the follicle. A few days following ovulation in the rabbit, as well as in other mammals, the entire surface of a ruptured follicle is covered by a superficial epithelium that is quite similar to that present in other areas of the ovary not affected by ovulation (Plate 34 and Chapter 1). The surface of the ovary that overlies a fully developed corpus luteum in the rabbit usually contains multicellular protrusions of varying size. For example, at about 6 to 7 days following ovulation, the apical areas of the corpus luteum may have numerous, flattened papillae that irregularly invaginate into the subjacent tissue (Plate 34). Both extensive evaginations (papillae) and invaginations (crypts) are characteristic of the rabbit ovary (see Chapter 1); therefore, it is not surprising to observe these structures on the surface of a corpus luteum. However, the organization and appearance of these papillae, in comparison to other areas of the ovary, may be a manifestation of the underlying organization of the luteal body. By contrast, the surface of the ovary that overlies the corpora lutea of the rat and mouse is generally free of papillae and other evaginations (*Van Blerkom* and *Motta,* 1978).

2.3.3. Electron Microscopic Observations on the Structure of the Corpus Luteum

While it is obvious that cellular movements are at work in the repair of the fractured area of an ovulated follicle, morphogenetic processes are also taking place within the interior of the follicle, although these processes are more difficult to observe directly with the scanning electron microscope. However, when viewed in cross section, the glandular organization of the corpus luteum is readily apparent by scanning electron microscopy (Plate 35 A). During the formation of a corpus luteum, luteal cells become arranged adjacent to capillary vessels in the form of cords or plates of cells. The cords of cells interconnect within the luteal mass, forming a labyrinthine or sponge-like structure.

During the initial stages of transformation into a corpus luteum, granulosa cells, mixed together with thin strands of connective tissue and numerous small vessels, hypertrophy, become polyhedral, and rapidly accumulate lipid (lutein) droplets. These modified cells are termed granulosa-lutein cells or, more simply, luteal cells. Results of numerous studies support the observation that luteal cells are derived primarily from granulosa cells, with some contribution from differentiated thecal or stromal cells (*Mossman* and *Duke,* 1973). Where luteal cells are not derived directly from granulosa cells, they are termed "paraluteal cells" or "theca-lutein cells" (see reviews of *Brambell,* 1956; *Watzka,* 1957; *Harrison,* 1962; *Mossman* and Duke, 1973; *Balboni,* 1976). Another putative cell type present in the early corpus luteum is the "K" cell, so called because of the high ketosteroid content of the cytoplasm (*White, Hertig, Rock,* and *Adams,* 1951). These cells appear to migrate from the theca interna to the granulosa layer of a recently ovulated follicle. As a historical footnote, it would seem

more appropriate to term these cells "Momigliano's cells" (if they need be named at all), after the investigator who first described them in 1927. At present, it is not clear whether "K" cells, "paraluteal," or "theca-lutein" cells are distinct, individual cell types or whether they simply represent transitional forms during the differentiation from a single, initial cell type (*Mossman* and *Duke,* 1973; *Balboni,* 1976). These cells have not been described by electron microscopy and are currently only identified by light microscopic techniques.

In contrast to the rather minor cellular components of the corpus luteum, the fine structure, histochemistry, and biochemistry of luteal cells have been examined extensively. In addition, no significant cellular or molecular differences have been observed when luteal cells derived from corpora lutea of pregnancy or corpora lutea of menstruation or even luteal cells from different species were compared (*Christensen* and *Gillim,* 1969; *Enders,* 1973). The "typical" luteal cell is a large, polyhedral element (20 to 40μm wide) containing numerous lipid droplets, mitochondria, and profiles of smooth-surfaced endoplasmic reticulum (Plate 36). The mitochondria possess an electron-dense matrix in which an amorphous material and/or granules are frequently observed (*Tokida,* 1965; *Motta,* 1969; *Adams* and *Hertig,* 1969a). The cristae contained within these mitochondria represent a variety of morphologies ranging from villiform and tubular to lamellar (*Enders* and *Lyons,* 1964; *Blanchette,* 1966a, b; *Priedkalns* and *Weber,* 1968; *Crisp, Dessouky,* and *Denys,* 1970; *Kurosumi* and *Fujita,* 1974) (Plates 36 B, 36 C). The numerous membranes of the smooth-surfaced endoplasmic reticulum form a complicated system of interconnecting, anastomosing tubules (Plates 35 B, 36 D), which often assume the unusual form of concentric whorls of membranes (*Blanchette,* 1966a, b; *Christensen* and *Gillim,* 1969; *Gillim, Christensen,* and *McLennan,* 1969). A Golgi complex is present in luteal cells, and Golgi vesicles are often observed in direct communication with elements of the smooth-surfaced endoplasmic reticulum (*Fawcett, Long,* and *Jones,* 1969; *Motta,* 1969; *Enders,* 1973; *Kurosumi* and *Fujita,* 1974).

Although the functional relationship among lipid droplets, mitochondria, and the endoplasmic reticulum in luteal cells is that characteristic of other steroidogenic cells (*Bjersing,* 1967a; *Christensen* and *Gillim,* 1969; *Enders,* 1973), the role of the Golgi complex, as well as the mechanism of secretion by luteal cells, is not completely understood. Several investigators have suggested that small, electron-dense bodies, sometimes associated with the Golgi, represent secretory granules (*Yamada* and *Ishikawa*, 1960; *Belt, Cavazos, Anderson,* and *Melampy*, 1971; *Gemmell, Stacy,* and *Thorburn*, 1974) released from the luteal cell cytoplasm by either an apocrine (*Yamada* and *Ishikawa*, 1960; *Green* and *Maqueo*, 1965) or ecrine mechanism (*Gemmell, Stacy,* and *Thorburn*, 1974). It is also generally accepted that the secretion of material produced by the luteal cells may involve the diffusion of molecular-sized particles across the plasma membrane (*Bjersing*, 1967b; *Christensen* and *Gillim*, 1969; *Fawcett, Long,* and *Jones*, 1969; *Motta*, 1969; *Crisp, Dessouky,* and *Denys*, 1970; *Enders*, 1973; *Kurosumi* and *Fujita*, 1974). Furthermore, it is known that the Golgi complex is involved in the synthesis of polysaccharide moieties of glycoproteins (*Enders*, 1973) which are ultimately incorporated into the luteal cell glycocalyx (*Familiari, Renda,* and *Motta*, 1978).

During the stages of active synthesis and secretion, the luteal cells of nearly all species examined contain both lysosome-like bodies (*Motta*, 1969; *Cohere, Brechenmacher,* and *Mayer*, 1967; *Vacek*, 1967; *Christensen* and *Gillim*, 1969; *Enders*, 1973) and (micro)peroxisomes (*Bock*, 1972; *Gulyas* and *Yuan*, 1975). In addition, microfilaments are quite evident both in granulosa cells in the process of differentiating into luteal cells as well as in mature luteal cells (*Adams* and *Hertig*, 1969a, 1969b; *Crisp, Dessouky,* and *Denys*, 1970; *Sinha, Seal,* and *Doe*, 1971; *Enders*, 1973; and *Motta* and *DiDio*, 1974; *Abel, Verhage, McClellan,* and *Niswender*, 1975). During the period of granulosa cell transformation, microfilaments are especially abundant in the cortical cytoplasm, where they may function in the morphogenetic reorganization of granulosa cells into a corpus luteum (*Motta* and *DiDio*, 1974). However, considering the distribution of microfilaments within the cytoplasm of a mature luteal cell, it is likely that contractile activity is highly localized in nature (*Enders*, 1973). Recent observations of the corpus luteum of the dog by high-voltage transmission electron microscopy demonstrate that microfilaments in mature luteal cells are arranged in a longitudinal fashion both between and around subcellular organelles, and that these microfilaments are directed toward pericapillary spaces (*Abel, McClellan, Verhage,* and *Niswender*, 1975). These observations of relatively thick sections suggest that microfilaments may be involved in the *compartmentalization* of organelles in the mature luteal cell (*Abel, Verhage, McClellan,* and *Niswender*, 1975).

The plasmalemma of a fully differentiated luteal cell is evaginated into both microvilli and larger cytoplasmic extensions which project into perivascular spaces (Plates 36B, 36C) (*Enders*, 1962; *Adams* and *Hertig*, 1969a, 1969b; *Sinha, Seal,* and *Doe*, 1971), as well as deeply into the cytoplasm of apposing cells (*Blanchette*, 1966b; *Gillim, Christensen,* and *McLennan*, 1969; *Motta*, 1969; *Abel, Verhage, McClellan,* and *Niswender*, 1975). As observed by scanning electron microscopy, the surface of a luteal cell contains numerous large, cytoplasmic projections, as well as corresponding and complementary cytoplasmic infoldings (Plate 37). The deep invaginations represent the regions of contact between the cytoplasmic projections of one luteal cell and the plasma membrane of another (luteal cells appear to become separated during preparation for scanning electron microscopy, thus permitting the direct observation of the cell surface). When the plasma membranes of luteal cells such as those described above are examined by high-resolution transmission electron microscopy, either with the use of tracers such as lanthanum nitrate or by freeze-fracture techniques, it is apparent that adjacent cells are in contact primarily by means of gap and septate junctions (*Albertini* and *Anderson*, 1975). Gap and tight junctions are also evident between granulosa cells in the preovulatory follicle (*Espey* and *Stutts*, 1972; *Albertini* and *Anderson*, 1974). Because it is currently held that gap junctions are involved in intercellular communication via ionic exchange (*Friend* and *Gilula*, 1972), it has been suggested that intercellular communication and coordination of cellular processes may be mediated by means of gap junctions both among granulosa and luteal cells (*Enders*, 1973; *Abel, Verhage, McClellan,* and *Niswender*, 1975; *Albertini* and *Anderson*, 1975; *Motta* and *Van Blerkom*, 1978). Support for the concept of facilitating intercellular coordination may be derived from the following electron microscopic observations: (1) the abundance and distribution of the cellular projections that are observed by scanning electron microscopy to infold deeply into the cytoplasm of adjacent cells (*Van Blerkom* and *Motta*, 1978), and (2) the nature of cellular contacts formed between the projections and the plasma membranes of apposing luteal cells, as demonstrated by transmission electron microscopy.

2.3.4. Regression of the Corpus Luteum

The functional life span of a corpus luteum is both species-dependent and also strictly related to whether fertilization has occurred. The termination of secretory activity by the luteal cells involves both the atrophy and degeneration of these elements as well as the involution of the entire luteal mass. At the fine structural level, regressing luteal cells contain autophagic vacuoles, heterogeneous lysosomes, large lipid inclusions, and myelin figures, which presumably are formed by the vacuolization of the membranous systems of the cytoplasm (*Carsten*, 1965; *Van Lennep* and *Madden*, 1965; *Adams* and *Hertig*, 1969a; *Motta*, 1969; *Bjersing, Hay, Moor*, and *Short*, 1970).

As observed during regression by scanning electron microscopy, the corpus luteum is a honeycomb-like structure in which each space is occupied by a partially vacuolized luteal cell having a mulberry-like appearance (Plate 38). Abundant connective tissue is present within the luteal mass and also forms a thick stroma that separates the degenerating luteal body from the surface of the ovary (Plate 38). The regression of the corpus luteum is not readily apparent from the surface of the ovary, since alterations within the luteal mass do not appear to affect overlying structures. The formation of a hyaline or fibrous body, termed the corpus albicans or fibrosus, is the final result of the complete degeneration and regression of the corpus luteum. While the corpora albicantia of a reproductive cycle in which pregnancy has not taken place generally shrink and disperse into the ovarian stroma rather rapidly, the complete regression of corpora lutea of pregnancy in some species may require several years before finally disappearing into the stroma (*Dubreuil and Riviere*, 1947).

2.4. The Luteo-Follicular Complex

On the basis of what has been presented both in this and the preceding chapter, it is clear that the basic cyclic structures of the ovary, which for traditional purposes are called the "follicle" and the "corpus luteum," are simply one structure that is continuously transforming under the stimulus of the pituitary gland. In this respect, this structure has been correctly identified since 1929, when it was termed the "apparato luteo-follicolare" (the luteo-follicular complex) (*Motta*, 1929, 1960). The apparent and extensive morphological dissimilarity that exists between the follicle and the corpus luteum is merely a manifestation of the different functions of this complex during the follicular (oocyte maturation) and luteal (secretion of hormones necessary for the maintenance of pregnancy) stages. In any event, these elements, with different structures and functions, are always derived from the same cells. Indeed, even ovulation, which so dramatically alters the morphology of this complex, does not interfere with the progressive transformation of the follicle into a corpus luteum – a process that is solely dependent upon the activity of the pituitary.

References

Abel, J. H., McClellan, M. C., Verhage, H. G., and *Niswender, G. N.* (1975). Subcellular compartmentalization of the luteal cell in the ovary of the dog. Cell Tiss. Res. *158*:461–480.

Abel, J.H., Verhage, H. G., McClellan, M. C., and *Niswender, G. N.* (1975). Ultrastructural analysis of the granulosa-luteal cell transition in the ovary of the dog. Cell. Tiss. Res. *160*:155–176.

Adams, E. C., and *Hertig, A. T.* (1969a). Studies on the human corpus luteum. I. Observations on the ultrastructure of development and regression of the luteal cells during the menstrual cycle. J. Cell Biol. *41*:696–715.

Adams, E. C., and *Hertig, A. T.* (1969b). Studies on the human corpus luteum. II. Observations on the ultrastructure of luteal cells during pregnancy. J. Cell Biol. *41*:716–735.

Albertini, D. F., and *Anderson, E.* (1974). The appearance and structure of intercellular connections during the ontogeny of rabbit ovarian follicles with particular reference to gap junctions. J. Cell Biol. *63*:234–250.

Albertini, D. F., and *Anderson, E.* (1975). Structural modifications of lutein cell gap junctions during pregnancy in the rat and mouse. Anat. Rec. *181*:171–194.

Albertini, D. F., Fawcett, D.W., and *Olds, P. J.* (1975). Morphological variations in gap junctions of ovarian granulosa cells. Tissue & Cell 7:389–405.

Amsterdam, A., Lindner, H., and *Groschel-Stewart, U.* (1977). Localization of actin and myosin in the rat oocyte and follicular wall by immunofluorescence. Anat. Rec. *187*:311–328.

Asdell, S. A. (1962). Mechanisms of ovulation. In: The Ovary, Vol. 2, p. 435 (*Zuckerman, S., Mandl, A. M.*, and *Eckstein, P.*). Academic Press, New York.

Balboni, G. C. (1976). Histology of the ovary. In: The Endocrine Function of the Human Ovary, pp. 1–24 (James, V. H. T., Serio, M., and Giusti, G., eds.). Proceed. Serono Symp., Vol. 7. Academic Press, New York.

Belt, W. D., Cavazos, K. F., Anderson, L. L., and *Melampy, R. M.* (1971). Cytoplasmic granules and relaxin levels in porcine corpora lutea. Endocrinology 89:1–10.

Bjersing, L. (1967a). On the ultrastructure of granulosa lutein cells in the porcine corpus luteum with special reference to endoplasmic reticulum and steroid hormone synthesis. Z. Zellf. mikrosk. Anat. *82*:187–211.

Bjersing, L. (1967b). Histochemical demonstration of Δ^5–3β- and 17β-hydroxysteroid dehydrogenase activities in porcine ovary. Histochemie *10*:295–304.

Bjersing, L., and Cajander, S. (1974a). Ovulation and the mechanism of follicle rupture. I, II, III. Cell Tiss. Res. *149*:287–327.

Bjersing, L., and Cajander, S. (1974b). Ovulation and the mechanism of follicle rupture. IV, V, VI. Cell Tiss. Res. *153*:1–99.

Bjersing, L., and Carstensen, H. (1964). The role of the granulosa cell in biosynthesis of ovarian steroid hormones. Biochim. Biophys. Acta (Amsterdam) *86*:639–640.

Bjersing, L., Hay, M. F., Moor, R. M., and Short, R. V. (1970). Endocrine activity, histochemistry and ultrastructure of ovine corpora lutea. I. Further observations on regression at the end of the oestrous cycle. Z. Zellforsch. mikrosk. Anat. *111*:437–457.

Björkman, N. (1962). A study of the ultrastructure of the granulosa cells of the rat ovary. Acta Anat. (Basel) *51*:125–147.

Blanchette, E. J. (1966a). Ovarian steroid cells. I. Differentiation of the lutein cells from the granulosa follicle cell during the preovulatory stage and under the influence of exogenous gonadotrophins. J. Cell Biol. *31*:501–516.

Blanchette, E. J. (1966b). Ovarian steroid cells. II. The lutein cell. J. Cell Biol. *31*:517–542.

Blandau, R. J. (1967). Anatomy of ovulation. Clin. Obstet. Gynecol. *10*:347–360.

Blandau, R. J. (1970). Growth of the ovarian follicle and ovulation. Prog. Gynec. *5*:58–76.

Bock, P. (1972). Peroxysomen im ovar der Maus. Z. Zellforsch. *133*:131–140.

Brambell, F. W. E. (1956). Ovarian changes. In: Marshall's Physiology of Reproduction, Vol. 1, Pt. 1 (Parkes, A. S., ed.). Longmans, Green and Co., London.

Burr, J. H., and Davies, J. I. (1951). The vascular system of the rabbit ovary and its relationship to ovulation. Anat. Rec. *111*:273–297.

Byskov, A. G. (1969). Ultrastructural studies on the preovulatory follicle in the mouse ovary. Z. Zellforsch. mikrosk. Anat. *100*:285–299.

Cajander, S. (1976). Structural alterations of rabbit ovarian follicles after mating with special reference to the overlying surface epithelium. Cell Tiss. Res. *173*:437–449.

Cajander, S., and Bjersing, L. (1975). Fine structural demonstration of acid phosphatase in rabbit germinal epithelium prior to induced ovulation. Cell Tiss. Res. *164*:279–289.

Cajander, S., and Bjersing, L. (1976). Further studies of the surface epithelium covering preovulatory rabbit follicles with special reference to lysosomal alterations. Cell Tiss. Res. *169*:129–141.

Caravaglios, R., and Cilotti, R. (1957). A study of the proteins in the follicular fluids of the cow. J. Endocrin. *15*:273–278.

Carsten, P. M. (1965). Elektronenmikroskopische Probleme bei Strukturdertungen von Einschlusskörpern im Menschlichen Corpus Luteum. Arch. Gynak. *200*:552–568.

Cavallotti, C., DiDio, L. J. A., Familiari, G., Fumagalli, G., and Motta, P. (1975). Microfilaments in granulosa cells of rabbit ovary: Immunological and ultrastructural observations. Acta Histochemica *52*:253–256.

Chang, S. C. S., Anderson, W., Lewis, W., Ryan, R. J., and Kang, Y. H. (1977). The porcine ovarian follicle. II. Electron microscopic study of surface features of granulosa cells at different stages of development. Biol. Repro. *16*:349–357.

Cherney, D. D., DiDio, L. J. A., and Motta, P. (1975). The development of rabbit ovarian follicles following copulation. Fertil. Steril. *26*:257–270.

Christensen, A. K., and Gillim, S. W. (1969). The correlation of fine structure and function in steroid secreting cells with emphasis on those of the gonads. In: The Gonads, pp. 415–488 (McKerns, K.W., ed.). Appleton, Century and Crofts, New York.

Christiansen, J. A., Jensen, C. E., and Zachariae, F. (1958). Studies on the mechanism of ovulation. Some remarks on the effects of depolymerization of high-polymers on the preovulatory growth of follicles. Acta Endocr. (Copenhagen) *29*:115–117.

Cohere, G., Brechenmacher, C., and Mayer, G. (1967). Variations des ultrastructures de la cellule luteale chez la ratte au cours de la grossesse. J. Microscopie (Paris) *6*:657–670.

Corner, G.W. (1919). On the origin of the corpus luteum of the sow from both granulosa and theca interna. Amer. J. Anat. *26*:117–183.

Corner, G.W., Jr. (1956). The histological dating of the human corpus luteum of menstruation. Amer. J. Anat. *98*:377–401.

Crisp, T. M., Dessouky, D. A., and Denys, F. R. (1970). The fine structure of the human corpus luteum of early pregnancy and during the progestational phase of the menstrual cycle. Amer. J. Anat. *127*:37–70.

Dubreuil, G., and Riviere, M. (1947). Morphologie et histologie des corps progestatifs et gestatifs (corps jaunes) de l'ovaire feminin. Gynecologie *43*:1–101.

Edwards, R. G., Fowler, R. E., Gore-Lanton, R. E., Gosden, R. G., Jones, E. C., Redhead, C., and Steptoe, P. C. (1977). Normal and abnormal follicular growth in mouse, rat and human ovaries. J. Repz. Fert. *51*:237–263.

Enders, A. C. (1962). Observations on the fine structure of lutein cells. J. Cell Biol. *12*:101–113.

Enders, A. C. (1973). Cytology of the corpus luteum. Biol. Reprod. *8*:158–182.

Enders, A. C., and Lyons, W. R. (1964). Observations on the fine structure of lutein cells. II. The effects of hypophysectomy and mammatrophic hormone in the rat. J. Cell Biol. *22*:127–141.

Espey, L. L. (1967). Ultrastructure of the apex of the rabbit Graafian follicle during the ovulatory process. Endocrinology *81*:267–276.

Espey, L. L. (1974). Ovarian proteolytic enzymes and ovulation. Biol. Reprod. *10*:216–235.

Espey, L. L., and Stutts, R. H. (1972). Exchange of cytoplasm between cells of the membrana granulosa in rabbit ovarian follicles. Biol. Reprod. *6*:168–175.

Familiari, G., Renda, T., and Motta, P. (1978). Surface coat in steroid secreting cells of the mouse ovary. Acta Anat. (Basel), *100*:193–202.

Fawcett, D. T., Long, J. A., and Jones, A. L. (1969). The ultrastructure of endocrine glands. Rec. Prog. Hormone Res. 25:315–280.

Friend, D. S., and Gilula, N. B. (1972). Variations in tight and gap junctions in mammalian tissues. J. Cell Biol. 53:758–776.

Fumagalli, Z., and Motta, P. (1969). Sulla presenza al microscopio elettronico di cellule muscolari lisce nell'ovaio di alcuni mammiferi. Atti. Soc. Ital. Anat. 28th Conv. Soc., Napoli.

Gemmel, R. T., Stacy, B. D., and Thorburn, G. D. (1974). Ultrastructural study of secretory granules in the corpus luteum of the sheep during the estrous cycle. Biol. Reprod. 11:447–462.

Gillim, S. W., Christensen, A. K., and McLennan, C. E. (1969). The fine structure of the human menstrual corpus luteum at its stage of maximum secretory activity. Amer. J. Anat. 126:409–428.

Green, J. A., and Maqueo, M. (1965). Ultrastructure of the human ovary. I. The luteal cell. Amer. J. Obstet. Gynecol. 92:946–957.

Gulyas, B. J., and Yuan, L. C. (1975). Microperoxisomes in the late pregnancy corpus luteum of rhesus monkeys. J. Histochem. Cytochem. 23:359–368.

Gwatkin, R. B. L. and Carter, H. W. (1975). Cumulus oophorus: In Scanning Electron Microscopic Atlas of Mammalin Reproduction. (E. S. E. Hafez, ed.) Springer Verlag, Berlin.

Hadjioloff, A. I., Bourneva, V., and Motta, P. (1973). Histochemical demonstration of Δ^5-3β-OHD activity in the granulosa cells of ovarian follicles of immature and mature mice correlated with some ultrastructural observations. Z. Zellforsch. mikrosk. Anat. 136:215–228.

Hafez, E. S. E. (1970). Reproduction and breeding techniques for laboratory animals. Lea and Febiger, Philadelphia.

Harrison, R. J. (1962). Ovarian structure (mammals). In: The Ovary, Vol. 1, pp. 143–187 (Zuckerman, S., Mandl, A. M., and Eckstein, P., eds.). Academic Press, New York.

Hertig, A.T., and Adams, E. C. (1967). Studies on the human oocyte and its follicle. Ultrastructural and cytochemical observations on the primordial follicle stage. J. Cell Biol. 34:647–675.

Hisaw, F. L. (1947). Development of the Graafian follicle and ovulation. Physiol. Rev. 27:95–119.

Kurosumi, K., and Fujita, H. (1974). An atlas of electron micrographs. Functional morphology of endocrine glands. Igaku Shoin Ltd., Tokyo.

McGaughey, R. W., and Van Blerkom, J. (1977). Patterns of polypeptide synthesis of porcine oocytes during maturation in vitro. Devel. Biol. 56:241–254.

Merk, F. B., Albright, J. T., and Botticelli, C. R. (1973). The fine structure of granulosa cell nexuses in rat ovarian follicles. Anat. Rec. 175:107–125.

Mestwerdt, W., Müller, O., and Brandau, H. (1977). Die differenzierte Struktur und Function der Granulosa und Theka in verschiedenen Follikelstadien menschlicher Ovarien. 1. Mitteilung: Der Primordial Primär-, Sekundär- und ruhende Tertiarfollikel. Arch. Gynak. 45–71.

Meyer, R. (1911). Über Corpus Luteum Bildung beim Menschen. Arch. Gynak. 93:354–404.

Momigliano, E. (1927). Sulla genesi del corpo luteo nella donna. Ric. Morf (Roma) 6:1–78.

Mossman, H .W., and Duke, K. L. (1973). Comparative Morphology of the Mammalian Ovary. University of Wisconsin Press, Madison, Wisc.

Motta, G. (1929). Sulla funzione dell'apparato luteofollicolare e sui rapporti tra il ciclo ovarico e quello endometriale. Arch. Ost. Gin. 16:260–341.

Motta, G. (1936). Beobachtungen und Betrachtungen über die Genese des Corpus luteum der Granulosa des follikels. Zentr. Gynak. 26:1547–1555.

Motta, G. (1960). Unita anatomic e funzionale del follicolo e del corpo luteo; l'apparato luteofollicaolare. Rass. Cl. Sc. 36:3–26.

Motta, P. (1965). Ricerche sulla formazione del "liquor folliculi" nell'ovaio della coniglia. Biol. Lat. 18:341–357.

Motta, P. (1969). Electron microscope study on the human lutein cell with special reference to its secretory activity. Z. Zellforsch. mikrosk. Anat. 98:233–245.

Motta, P., Cherney, D. D., and DiDio, L. J. A. (1971). Scanning and transmission electron microscopy of the ovarian surface in mammals with special reference to ovulation. J. Submicr. Cytol. 3:85–100.

Motta, P., Takeva, S., and Nesci, E. (1971). Etude ultrastructurale et histochemique des rapports entre les cellules folliculaires et l'ovocyte pendant le developement du follicule ovarien chez les mammiferes. Acta Anat. (Basel) 80:537–562.

Motta, P., and DiDio, L. J. A. (1974). Microfilaments in granulosa cells during the development of the follicle and its transformation into a corpus luteum in the rabbit ovary. J. Submicr. Cytol. 6:15–27.

Motta, P., and Van Blerkom, J. (1974). A scanning electron microscopic study of the luteo-follicular complex. I. Follicle and oocyte. J. Submicr. Cytol. 6:297–310.

Motta, P., and Van Blerkom, J. (1975). A scanning electron microscoepic study of the luteo-follicular complex. II. Events leading to ovulation. Amer. J. Anat. 143:241–264.

Motta, P., and Van Blerkom, J. (1978). Structure and ultrastructure of the Graafian follicle. In: Human Ovulation: Mechanisms, Prediction, Detection and Regulation (Hafez, E. S. E., ed.). North Holland, The Nederlands.

Nilsson, O., and Munshi, S. F. (1973). Scanning electron microscopy of mouse follicles at ovulation. J. Submicr. Cytol. 5:1–6.

Odor, D.L. (1960). Electron microscopic studies on ovarian oocytes and unfertilized tubal ova in the rat. J. Biophys. Biochem. Cytol. 7:567–574.

Odor, D. L., and Blandau, R. J. (1951). Observations on the formation of the second polar body in the rat ovum. Anat. Rec. 110:329–347.

Odor, D. L., and Blandau, R. J. (1973). Egg transport over the fimbrial surface of the rabbit oviduct under experimental conditions. Fertil. Steril. 24:292–300.

O'Shea, J. D. (1970). An ultrastructural study of smooth muscle-like cells in the theca externa of ovarian follicles in the rat. Anat. Rec. 167:127–140.

Osvaldo-Decima, L. (1970). Smooth muscle in the ovary of the rat and monkey. J. Ultrastruct. Res. 30:218–237.

Parr, E. L. (1974). Histological examination of the rat ovarian follicle wall prior to ovulation. Biol. Reprod. 11:483–503.

Pratt, J. P. (1935). The human corpus luteum. Arch. Path. Lab. Med. 19:380–545.

Priedkalns, J., and Weber, A. F. (1968). Ultrastructural studies of the bovine Graafian follicle and corpus luteum. Z. Zellforsch. mikrosk. Anat. 91:554–573.

Rawson, J. M. R. and Espey, L. L. (1977). Concentration of electron dense granules in the rabbit ovarian surface epithelium during ovulation. Biol. Reprod. 17:561–566.

Rona, G. (1963). The role of vascular mucopolysaccharides in the hemostatic action of estrogens. Am. J. Obst. Gynec. 87:434–444.

Rondell, P. (1970). Biophysical aspects of ovulation. Biol. Reprod. (suppl.) 2:64–89.

Rosenbauer, K. A., Jansen, B., and Lindauer, S. (1976). Rasterelektronenmikroskopische Analyse der Eierstockoberfläche und ihrer Veränderungen im Zusammenhang mit dem Ovulationsvorgang. GiT Fachzeit. f. Labor. 4:325–329.

Schultz, R.M., and Wassarman, P. M. (1977). Specific changes in the pattern of protein synthesis during meiotic maturation of mammalian oocytes in vitro. Proc. Nat. Acad. Sci. (USA) 74:538–541.

Sinha, A. A., Seal, U. S., and Doe, R. P. (1971). Fine structure of the corpus luteum of the raccoon during pregnancy. Z. Zellforsch. mikrosk. Anat. 117:35–45.

Sotelo, J. R., and Porter, K. R. (1959). An electron microscope study of the rat ovum. J. Biophys. Biochem. Cytol. 5:327–341.

Szego, C. M., and Gitin, E. S. (1964). Ovarian histamine depletion during acute hyperemic response to luteinizing hormone. Nature (London) 201:682–684.

Talo, A. (1974). Electric and mechanical activity of the rabbit oviduct in vitro before and after ovulation. Biol. Reprod. 11:335–345.

Tokida, A. (1965). Electron microscopic studies of the corporal lutea obtained from normal human ovaries. Mie. Med. J. 15:27–76.

Vacek, Z. (1967). Ultrastructure and enzyme histochemistry of the corpus luteum graviditatis and its correlation to the decidual transformation of the endometrium. Folia Morph. 15:375–383.

Van Blerkom, J. (1977). Molecular approaches to the study of oocyte maturation and embryonic development. In: Immunobiology of the Gametes, pp. 187–206 (Edidin, M., and Johnson, M. H., eds.). Cambridge University Press, Cambridge, England.

Van Blerkom, J., and McGaughey, R.W. (1978). Molecular differentiation of the rabbit ovum. I. During oocyte maturation in vivo and in vitro. Devel. Biol. 63:

Van Blerkom, J., and Motta, P. (1978). A scanning electron microscopic study of the luteo-follicular complex. III. Formation of the corpus luteum and repair of the ovulated follicle. Cell Tiss. Res. 189:131–154.

Van Lennep, E. W., and Madden, L. M. (1965). Electron microscopic observations on the involution of the human corpus luteum of menstruation. Z. Zellforsch. mikrosk. Anat. 66:365–380.

Walles, B., Edvinsson, L., Owman, C., Sjoberg, N. O., and Sporrong, B. (1976). Cholinergic nerves and receptors mediating contraction of the Graafian follicle. Biol. Reprod. 15:565–572.

Warnes, G. M., Moor, R. M., and Johnson, M. H. (1977). Changes in protein synthesis during maturation of sheep oocytes in vitro and in vivo. J. Reprod. Fert. 49:331–335.

Wassarman, P. M., and Letourneau, G. E. (1976). RNA synthesis in fully grown mouse oocytes. Nature (London) 261:73–74.

Watzka, A. M. (1957). Weibliche Genitalorgane. Das Ovarium. In: Handbuch der Mikroskopischen Anatomie des Menschen, Vol. 7, pp. 1–178 (Mollendorf and Bargmann, eds.). Springer-Verlag, Berlin.

Weir, B. (1977). Ovulation: 19th Symposium of Soc. Fert. (Sheffield). J. Reprod Fert. 51:179–264.

White, R. F., Hertig, A. T., Rock, J., and Adams, E. C. (1951). Histological and histochemical observations on the corpus luteum of human pregnancy with special reference to corpora lutea associated with early normal and abnormal ova. Contr. Embryol. Carneg. Inst. 36:55–74.

Yamada, E., and Ishikawa, T. M. (1960). The fine structure of the corpus luteum in the mouse ovary as revealed by electron microscopy. Kyusku J. Med. Sci. 11:235–259.

Yang, N. S. T., Marsh, J. M., and LeMaire, W. J. (1974). Post-ovulatory changes in the concentration of prostaglandins in rabbit Graafian follicles. Prostoglandins 6:37–44.

Zamboni, L. (1970). Ultrastructure of mammalian oocytes and ova. Biol. Reprod. 2 (suppl.):44–63.

Zamboni, L. (1971). The Fine Morphology of Mammalian Fertilization. Harper & Row, New York.

Zamboni, L., and Mastroianni, L., Jr. (1966). Electron microscopic studies on rabbit ova. I. The follicular oocyte. J. Ultrastruct. Res. 14:95–117.

Plate 18. The Preovulatory Follicle.

A   Light micrograph of a preovulatory follicle in which the antrum has been distended as a result of the accumulation of liquor folliculi (Lf). Granulosa cells (Gc) are generally dissociated from one another and the oocyte appears to be almost free-floating within the antrum (arrow*). The stratum granulosum is slightly plicated (arrow), and within the theca (Ti), patches of epithelioid cells (thecal gland) (*) are evident. (x 65; rabbit, 6 hours postcoitum).

B   Light micrograph of a preovulatory follicle in which some granulosa cells (Gc, arrows) appear to be undergoing precocious luteinization as is indicated by the presence of dark bodies in the cytoplasm. The vessels located within the theca interna (Ti) contain blood cells (hemostasis) (arrows), while in intercellular areas, fluids have accumulated (edema) (arrows*). Lf = liquor folliculi. (x 550; rabbit, 8 hours postcoitum).

C   Light micrograph of an oocyte and surrounding coronal cells that have become detached from the granulosa wall just prior to ovulation. Note the striations in the zona pellucida (Zp) which represent the ameboid processes of the coronal cells (Gc). The relatively prominent structures in the perivitelline space correspond to the terminal portions of the processes where the plasma membranes of the oocyte and granulosa cells are in contact by means of specialized junctions (arrows). All sections for light microscopy were 1 μm thick and were stained with toluidine blue. (x 950; rabbit, 10 hours postcoitum).

66

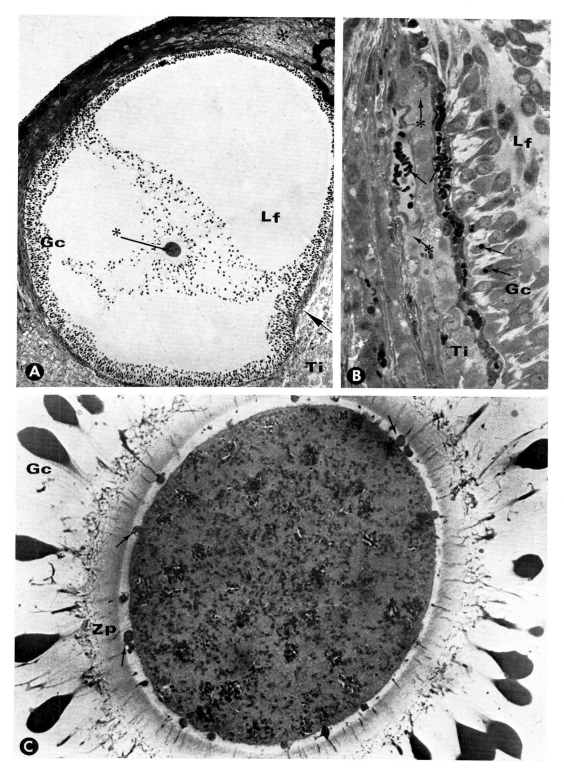

Plate 18. The Preovulatory Follicle.

Plate 19. Preovulatory Follicle: Granulosa and Theca.

A   Transmission electron micrograph illustrating the appearance of partially luteinized granulosa cells (Gc), as well as the dissociation of the theca interna (Ti) and tunica albuginea (Ta) by fluids composed of liquor folliculi (Lf) and other intercellular fluids (*). Cytolysis of the cells of the theca and tunica albuginea is also evident in this subapical area. (x 3,900; rabbit, 8 hours postcoitum).

B   Transmission electron micrograph demonstrating the complete dissociation of the tunica albuginea (Ta) in a subapical zone. The disruption is apparently caused by the accumulation of fluids (edema, *) in which are evident blood cells (primarily leukocytes, Le) that are liberated from the capillaries after the rupture of the endothelial wall. (x 9,750; rabbit, 10 hours postcoitum).

C   High-magnification transmission electron micrograph of a preovulatory follicle in which granulosa cells are in the process of undergoing luteinization. Present within the cytoplasm of this particular cell are bundles of microfilaments (Mf), lipid droplets (L), a Golgi complex (G), and mitochondria (M). Lf = liquor folliculi. (x 9,700; rabbit, 2 hours postcoitum).

D   Transmission electron micrograph of the evaginations of granulosa cells in a preovulatory follicle. Note both regions of intercellular contact (arrows*) and clusters of microfilaments (Mf). (x 9,100; rabbit, 2 hours postcoitum).

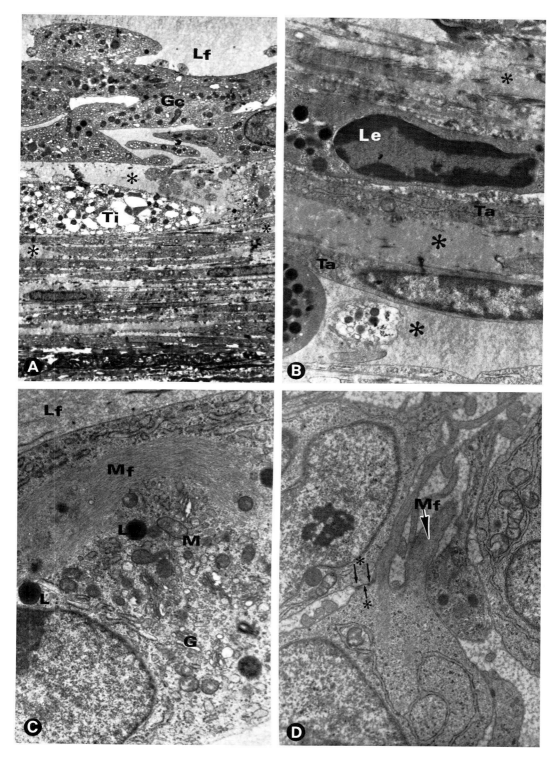

Plate 19. Preovulatory Follicle: Granulosa and Theca.

Plate 20. Preovulatory Follicle: Tunica Albuginea and Supergicial Epithelium.

A   This transmission electron micrograph illustrates the appearance of the apical wall of a preovulatory follicle. A fluid-like material has accumulated in the connective tissue of the tunica albuginea (Ta), below the
basal lamina (Bl), and in the spaces between the cells of the superficial epithelium (Se). This material is so
abundant in some regions as to cause extensive swelling of intercellular spaces (*). A few microvilli are
evident on the surface of the superficial cells (Mv), and present within the cytoplasm are spheroidal bodies
(d) that may correspond to lysosomes and/or small fluid-filled vesicles. (x 7,000; rabbit, 10 hours postcoitum).

B   At the apical wall of a preovulatory follicle, the superficial epithelium (Se) has already been shed, leaving
behind cellular debris. The connective matrix of the tunica albuginea (Ta) consists of dissociated fibers,
degenerating cells, and "pools" of a fluid-like material (*). (x 13,500; rabbit, 10 hours postcoitum).

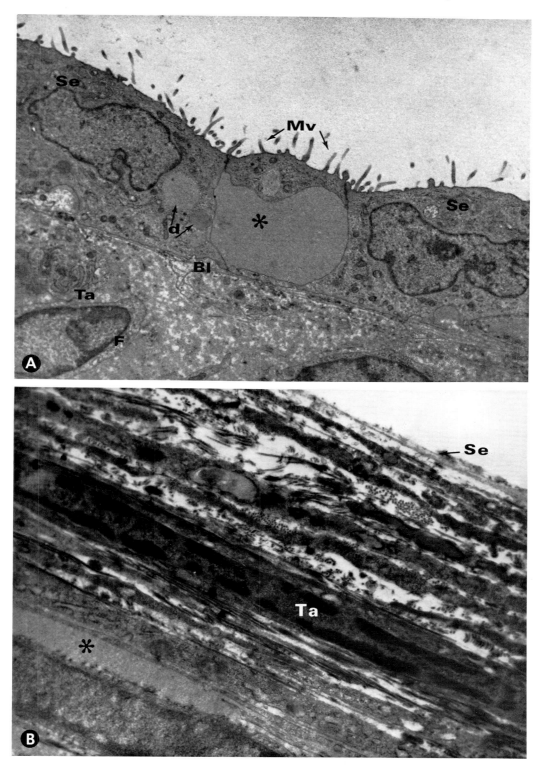

Plate 20. Preovulatory Follicle: Tunica Albuginea and Superficial Epithelium.

Plate 21. The Surface of a Preovulatory Follicle.

A   As observed by scanning electron microscopy, papillae (P) are frequently arranged in a flower-like manner around the protruding surface of a rabbit preovulatory follicle. (x 94; 8 hours postcoitum).

B   Scanning electron micrograph demonstrating the appearance of extremely flattened cells that populate
the apex of a preovulatory follicle (arrow). While the cells covering the lateral surfaces are elongated (arrow 1), the cells at the base of the follicle are polyhedral (arrow 2). (x 315; mouse in late proestrous).

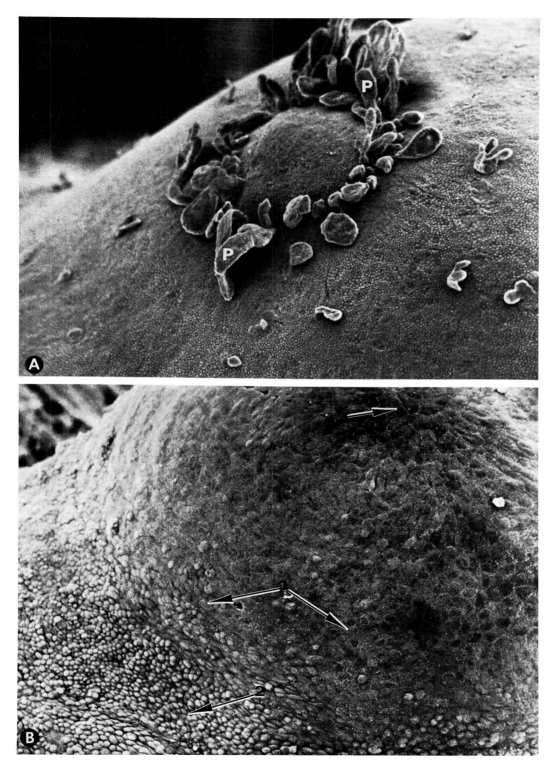

Plate 21. The Surface of a Preovulatory Follicle.

Plate 22. Preovulatory Follicles.

A   A preovulatory follicle protruding from the surface of a rat ovary is evident in this scanning electron micrograph. Note the appearance of the superficial epithelium covering the basal (1), lateral (2), and apical (3) aspect of the follicle. (x 162).

B   The apex (*) of a rabbit preovulatory follicle at 9 hours postcoitum is not completely covered by the superficial epithelium (Se). In other regions, patches of the superficial epithelium are in the process of sloughing off (arrows). The area of the apex that protrudes above surrounding zones (delineated by a dotted line) most likely corresponds to the "stigma." (x 120).

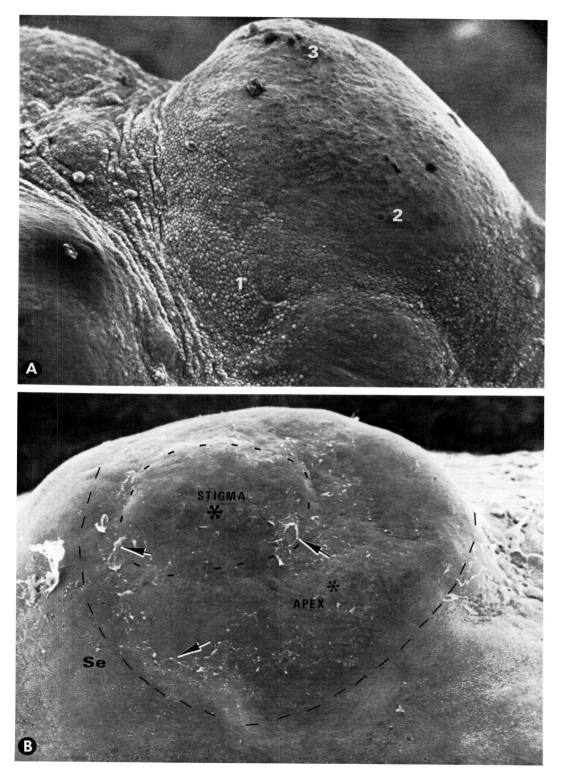

Plate 22. Preovulatory Follicles.

Plate 23. Preovulatory Follicles.

A The protruding apex of a rabbit preovulatory follicle at 10 hours postcoitum is shown in this scanning electron micrograph. Note the complete loss of the superficial epithelium in some regions and the sloughing off of superficial cells in other apical areas (arrows). (x 33).

B An additional view of the follicle shown in A, demonstrating the extensive loss of the superficial epithelium. This surface appearance is characteristic of rabbit follicles in the period immediately preceding ovulation (the rabbit usually ovulates at approximately 12 hours postcoitum). The arrow indicates a mass of cells and possibly fluids which is frequently observed on the surface of preovulatory rabbit follicles. (x 53).

C Scanning electron micrograph of a fractured preparation of a large, cavitary follicle at the same preovulatory stage as the follicles shown in A and B above. The follicular antrum contains liquor folliculi (Lf) and small groups of cells belonging to the cumulus oophorus (arrow). The granulosa layer is thinned and slightly plicated in some regions (arrows*). This micrograph should be compared to Plate 18 A since extensive clusters of interstitial cells (Ic) (interstitial glands) are present in the perifollicular regions. Se = superficial epithelium, P = papillae. (x 88; rabbit, 10 hours postcoitum).

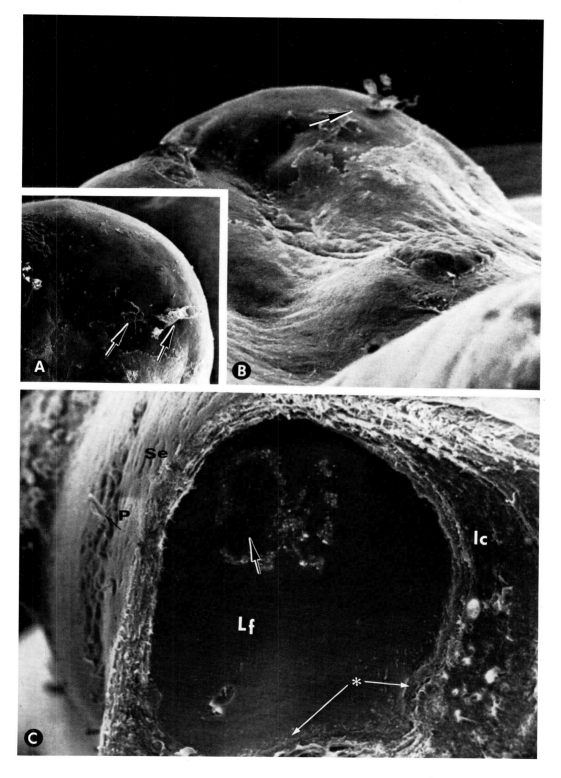

Plate 23. Preovulatory Follicles.

Plate 24. The Superficial Epithelium of Preovulatory Follicles.

A   Scanning electron micrograph demonstrating the characteristic appearance of superficial cells in the basal and more lateral areas of a preovulatory rabbit follicle at 10 hours postcoitum. These cells are polyhedral elements which are covered with microvilli and express isolated cilia (arrow). (x 4,200).

B   Scanning electron micrograph illustrating the surface topology of a preovulatory follicle in a rabbit at 11 hours postcoitum. Underlying fluids appear to have emerged onto the surface of the remaining superficial epithelium (Se) in the form of blebs and/or free droplets (arrows). The remaining superficial cells in the apical region are quite flattened and relatively devoid of microvilli. This morphology is in sharp contrast to the appearance of superficial cells in basal and lateral regions as described above in A. (x 4,600).

C   Scanning electron micrograph depicting the disruption of the superficial cells that eventually leads to their sloughing off from the apex of a follicle. (x 2,460; rabbit, 9 hours postcoitum).

D   Scanning electron micrograph of an area near the apex of a rabbit preovulatory follicle at 10 hours postcoitum. The extensive disruption of the superficial epithelium (Se) has resulted in the exposure of the subjacent connective tissue of the tunica albuginea (Ta). (x 650).

78

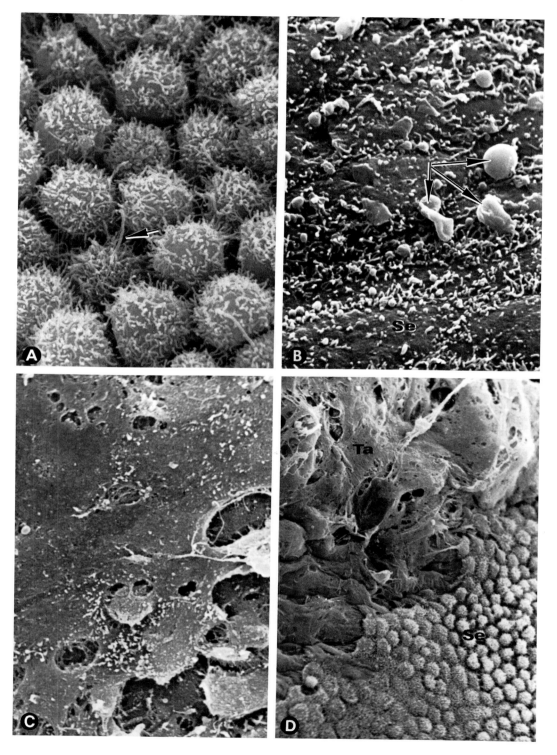

Plate 24. The Superficial Epithelium of Preovulatory Follicles.

Plate 25. Ovulatory Follicles.

A   Scanning electron microscopic stereo view of an oocyte (*), most probably in the process of ovulation. The oocyte, immersed in a large quantity of fluid and cytoplasmic debris, appears to emerge from within the follicular cavity through a rupture in the apex of the follicular wall. (x 230; estrous mouse).

B   Scanning electron microscopic stereo view of the apex of a follicle just after ovulation in the rabbit. From the surface, it is possible to observe directly the ruptured stigma that is surrounded by a gel-like mass of follicular fluids and clusters of granulosa cells. The collapsed residual cavity of the follicle is clearly evident from the surface (*). In the lower right-hand region of the micrograph, an oocyte, covered by the cells of the corona radiata (Cr), is evident. (x 100).

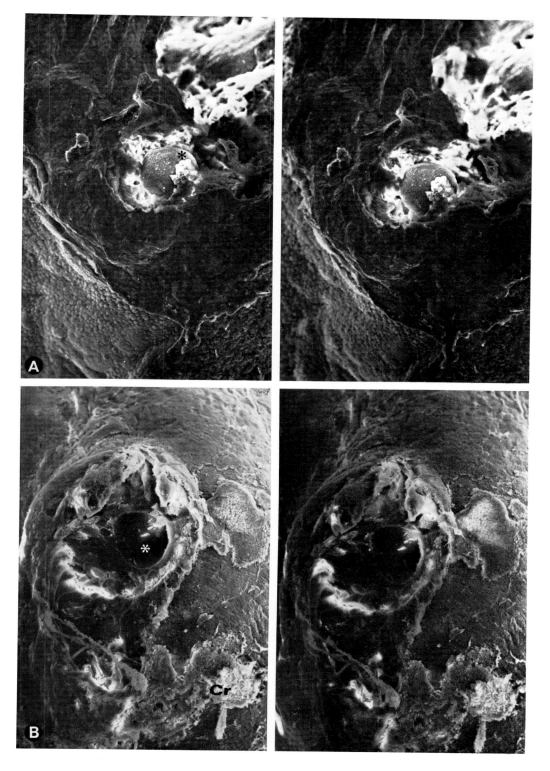

Plate 25. Ovulatory Follicles.

Plate 26. Ovulation.

It quite likely that this scanning electron micrograph demonstrates the actual process of ovulation in the mouse. The surface of the oocyte (Oo) is covered by the zona pellucida (Zp), which is itself only partially surrounded by granulosa cells (Gc) (some of which may have been removed during preparation of specimen for scanning electron microscopy). (x 2,300).

Plate 26. Ovulation.

Plate 27. Ovulated Follicle and Cumulus Mass.

In this scanning electron micrograph of a rabbit follicle between 30 minutes and 1 hour following ovulation, numerous granulosa cells (Gc), immersed in a coagulated fluid, cover regions adjacent to the ruptured apex. The oocyte (arrow, Oo), enclosed by a mass of coronal cells and follicular fluids (Lf), is evident near the site of ovulation (*). (x 225).

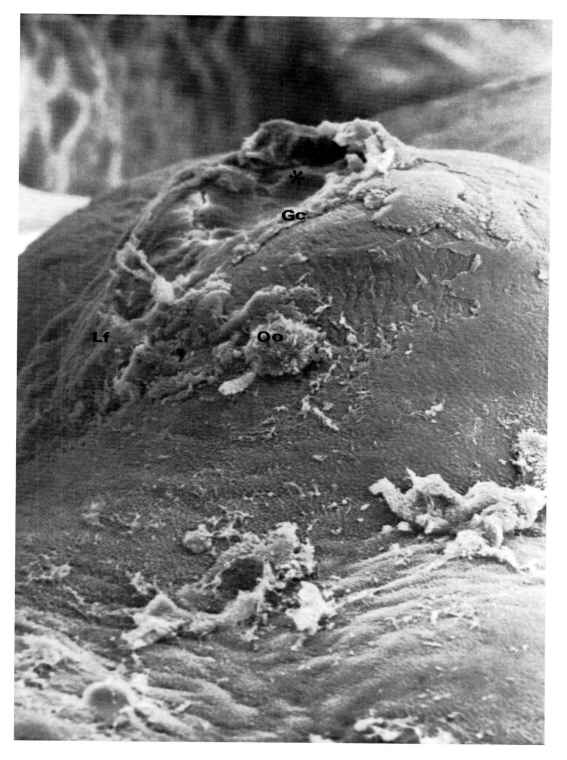

Plate 27. Ovulated Follicle and Cumulus Mass.

Plate 28. Postovulatory Follicle.

A   The surface of a rabbit follicle at approximately 9 hours postovulation is shown in this scanning electron micrograph. A mass of amorphous material (*) protrudes from the ruptured surface and the surrounding areas are covered by a distorted and discontinuous layer of superficial cells (Se). (x 264).

B   Scanning electron microscopic stereo view of a rabbit follicle at approximately 4 hours following ovulation. A mass of follicular fluids and coagulated plasma, to which are attached residual cells, protrudes markedly above the surface of the ovary. The formation of this mass probably represents the initial stage in the repair of the ruptured follicle. (x 100).

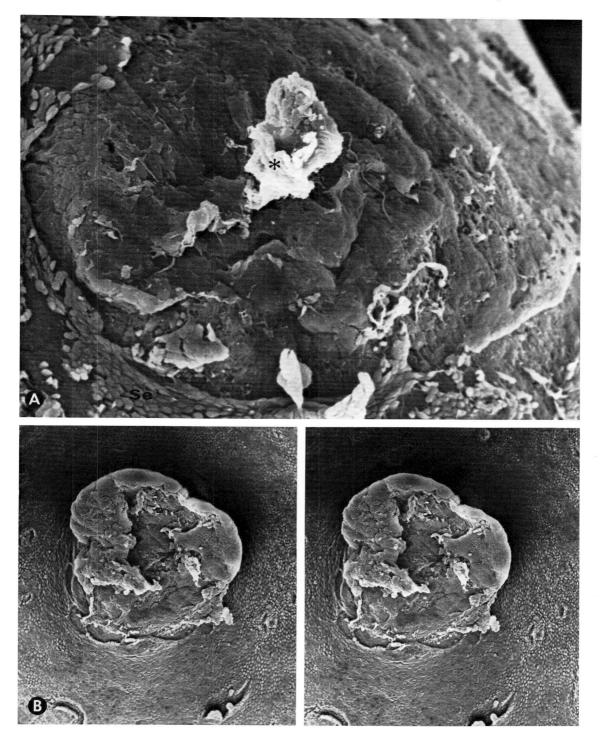

Plate 28. Postovulatory Follicle.

Plate 29. Postovulatory Follicles.

A   The ruptured apex of a rabbit follicle at approximately 6 hours following ovulation is shown in this scanning electron micrograph. An amorphous, gel-like material (*) is evident in the center of the follicle, while the ruptured zone is occupied by numerous plications of the granulosa wall (Gc). The superficial epithelium (Se) at the basal areas of the follicle appears to be undisturbed. (x 360).

B   This comparatively low-magnification scanning electron micrograph of a rabbit ovary estimated at 1.5 days following ovulation demonstrates the appearance of several ovulated follicles. Ovulated follicles protude irregularly above the surface of the ovary and each follicle is covered by a rather viscous, gel-like material (arrows). This material is most probably composed of residual cells and cellular debris (originating from the follicular wall and the connective tissue of the tunica albuginea) and coagulated liquor folliculi and plasma. (x 25).

C   Scanning electron micrograph of a rabbit follicle estimated at 1.5 days following ovulation. This micrograph demonstrates the rather copious amount of fluid (coagulated) which is discharged at ovulation but, in this infrequent case, is still attached to the apical zone. P = papillae. (x 58).

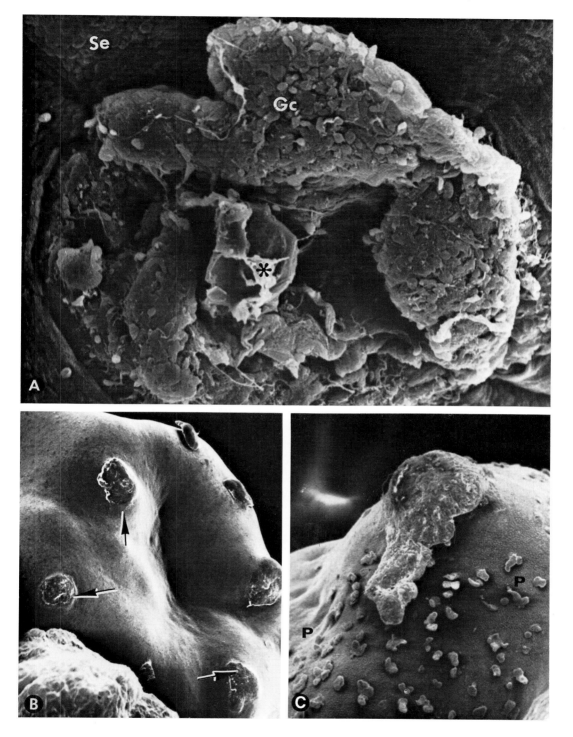

Plate 29. Postovulatory Follicles.

Plate 30. Postovulatory Follicles.

A   Between approximately 1 and 1.5 days following ovulation in the rabbit, the area of rupture is occupied by a mass of connective tissue, to which residual blood and granulosa cells are still attached. In this view, the region of disruption caused by ovulation is clearly separated from the relatively undisturbed, basal, superficial epithelium (Se). (x 280).
Insert: Light micrograph of a rabbit follicle quite similar to that observed by scanning electron microscopy in A.

B   Scanning electron micrograph of the protruding apex of an ovulated rabbit follicle between 1.5 and 2 days following ovulation. Groups of superficial cells (Se) have begun to proliferate and to migrate (arrows) over the subjacent connective tissue stroma (Ct) that seals the site of dehiscence. Note in particular how the connective cells (primarily fibroblasts, *) protrude through this extremely thin superficial layer. (x 470).

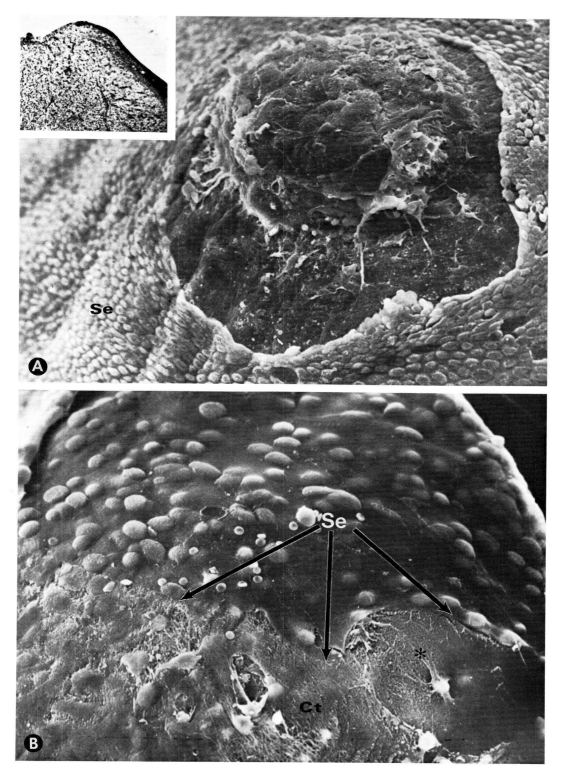

Plate 30. Postovulatory Follicles.

Plate 31. Postovulatory Rabbit Follicles.

A   Scanning electron microscopic view of the boundary between migrating superficial cells (Se) and the subjacent connective tissue (Ct). The bulges (arrows) in the flattened layer of superficial cells probably correspond to the cellular bodies of underlying fibroblast-like cells, or to the nuclei of the superficial cells. (x 1,090; 1.5 to 2 days following ovulation).

B   High-magnification scanning electron micrograph of a region of connective tissue similar to that shown in A (arrows *). Note the large accumulation of fibers surrounding a connective cell (arrow). (x 1,820; 1.5 to 2 days following ovulation).

C   Higher magnification view of the laminar and extremely flattened superficial cells that are in the process of migrating over the subjacent connective tissue mass. The protruding zones correspond to the cell bodies of underlying connective elements (*) (or possibly the nuclei of the superficial cells). An area of discontinuous cell contact ("zipper-like") is characteristic of these cells and is evident between the laminar borders of two adjacent elements (arrows). (x 2,860; 1.5 to 2 days following ovulation).

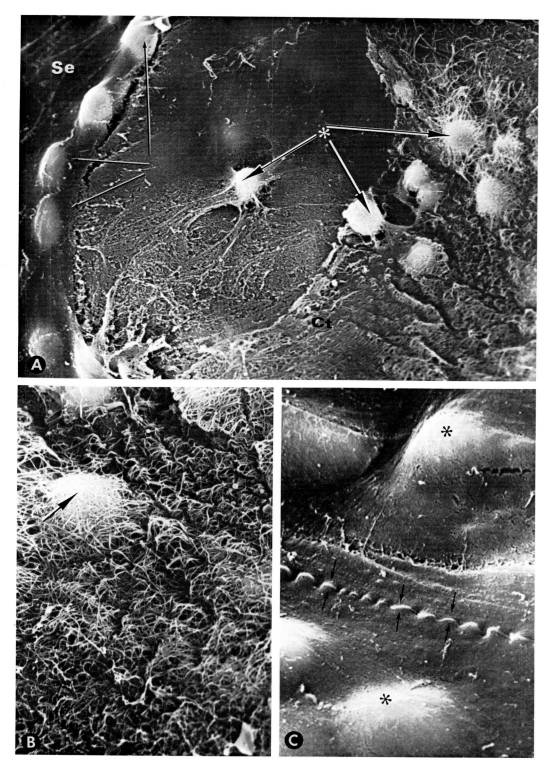

Plate 31. Postovulatory Rabbit Follicles.

Plate 32. Postovulatory Follicles.

A   Scanning electron micrograph of the apical region of a rabbit follicle at approximately 1.5 days following ovulation. The presence of numerous connective cells (Ct) on the surface of the residual mass of coagulated fluids indicates that this follicle is in an early stage of repair. The boundary between the superficial epithelium (Se) and the disrupted portion of the follicle is clearly evident. Numerous cells of the basal superficial epithelium exhibit prominent apical protrusions (arrows). (x 630).
Insert: Light micrograph of an ovulated rabbit follicle at the same stage as the follicle shown in A. The arrow indicates a mass of coagulated fluids that are discharged at ovulation.

B   Scanning electron micrograph of cellular protrusions in a region of the subapical, superficial epithelium (arrows). (x 3,000; rabbit, 1.5 days following ovulation).

C   Higher-magnification scanning electron microscopic view of the bulbous projections (arrows) of some superficial cells located in the subapical region of a rabbit follicle at approximately 1.5 days following ovulation. The functional significance of these cellular projections is unknown, although they may possibly be related to a process of endocytosis. (x 5,800).

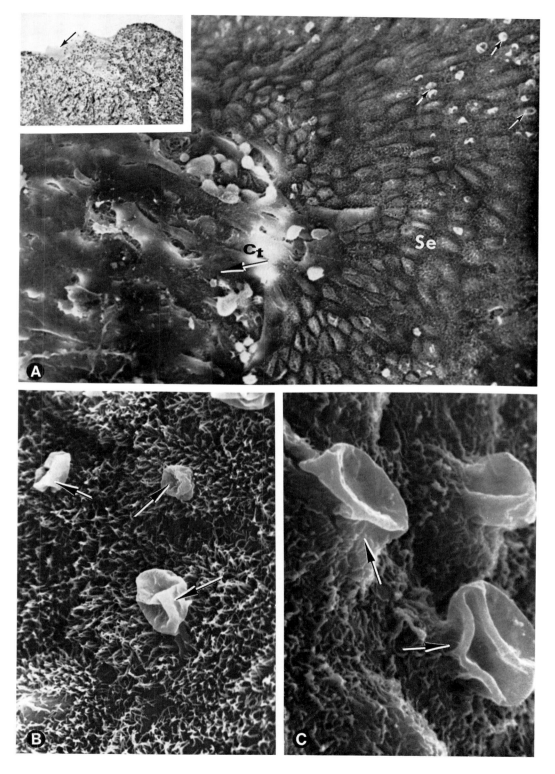

Plate 32. Postovulatory Follicles.

Plate 33. Postovulatory Rabbit Follicles.

A   The apex of a postovulatory follicle (the luteo-follicular complex) still extends markedly above the surface of the ovary at approximately 2 days following ovulation. In evidence are both connective tissue covering the apex (Ct) and areas of proliferating and migrating superficial cells originating from the more basal areas of the follicle (Se). (x 203).

B   At higher magnification, cells of the superficial epithelium present at the basal areas of the follicle appear to "flow" toward the disrupted apical regions. Transition forms of superficial cells, from polyhedral to extremely flattened, are visible as these elements migrate over the connective tissue matrix (arrows). The region of the follicle from which this view of superficial cells (Se) originated is outlined in A. (x 640).

C   High-magnification scanning electron micrograph of flattened superficial cells (Se) in the process of migrating over the surface of the connective tissue (Ct). Cellular contacts (arrows) and microvilli (Mv) are evident on the surfaces of these cells. (x 3,000; 2 days following ovulation).

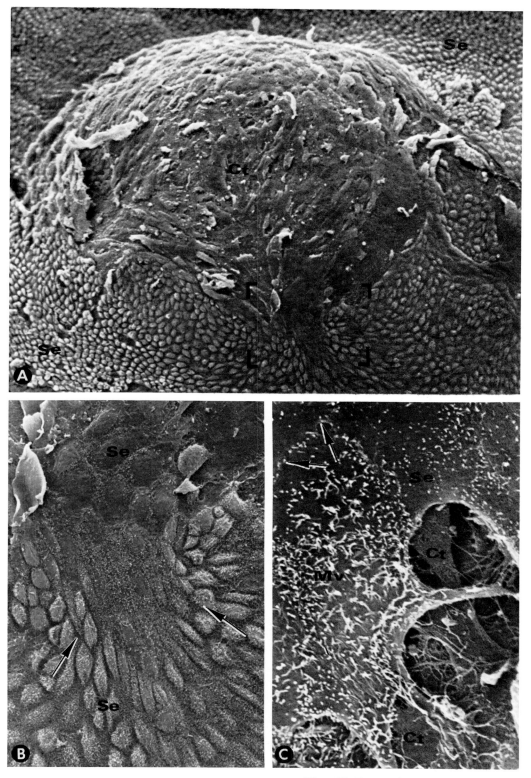

Plate 33. Postovulatory Rabbit Follicles.

Plate 34. Surface of the Corpus Luteum.

A   The appearance of the ovarian surface that overlies a corpus luteum (Cl) at 7.5 days following ovulation is depicted in this scanning electron micrograph. The superficial epithelium (Se) has completed proliferation and now covers the entire site of ovulation. In addition, this region of the ovarian surface is composed of numerous plicae and flattened papillae (P) which invaginate in an irregular fashion into the body of the corpus luteum. (x 130; rabbit, 8 days postcoitum).
Insert: A light microscopic view of the surface of a corpus luteum similar to that represented in A. Note the papilla-like evagination (arrow).

B   High magnification of A. Numerous papillae (P) and cords of cells are not only quite flattened but also appear to infold into subjacent areas of the corpus luteum. These surface structures may be a manifestation of hormone production by the luteal cells or simply an expression of the proliferative activity of the superficial epithelium (see Chapter 1). These features are, however, characteristic of rabbit corpora lutea. (x 385).

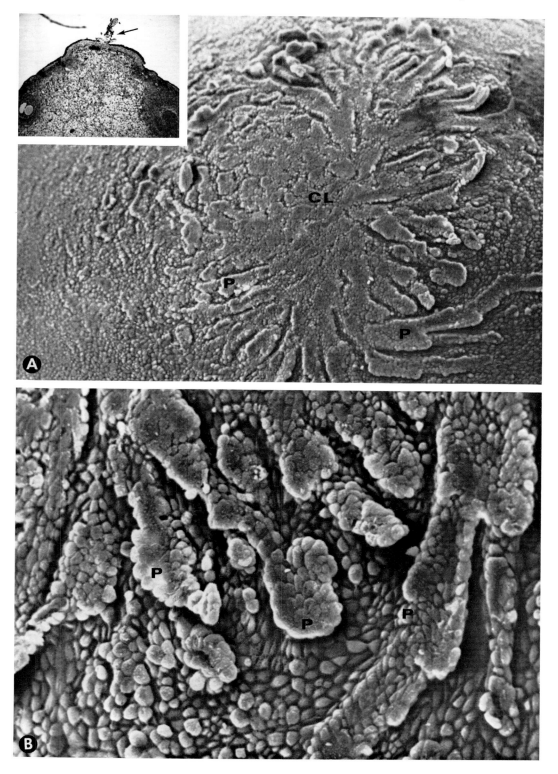

Plate 34. Surface of the Corpus Luteum.

Plate 35. The Corpus Luteum.

A   Scanning electron micrograph of the cut surface of a mature corpus luteum in the rat. Luteal cells are arranged in irregular cords that are surrounded by numerous capillaries (Ca). The periphery of the luteal body is covered by a connective tissue capsule (Ct) containing large vessels (V). Other vessels (V$_1$) are evident in the interior region of the luteal body. (x 165; rat, midpregnancy).

B   Scanning electron microscopic stereo view of the surface of a section through a mature luteal cell. N = nucleus, Nm = nuclear membrane. The numerous anastomosing structures within the cytoplasm likely represent tubules of the smooth-surfaced endoplasmic reticulum (*). (x 12,500; rat, midpregnancy).

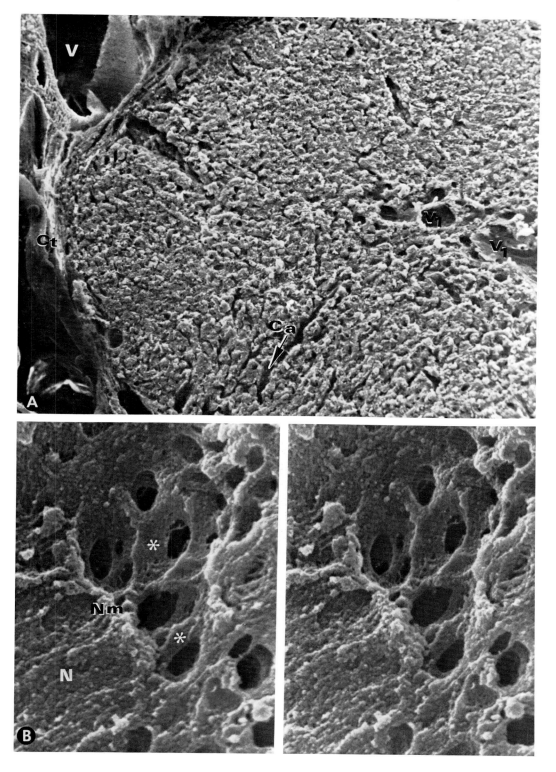

Plate 35. The Corpus Luteum.

Plate 36. The Corpus Luteum.

A Light micrograph of a portion of a mature corpus luteum. Granulosa-luteal cells = Lc; theca lutein cells (or paraluteal cells) = Tlc; capsule of connective tissue surrounding the corpus luteum = Ct. (5 micron section stained with Azan Mallory stain). (x 230; rat).

B Transmission electron micrograph of cords of differentiated luteal cells. Mitochondria (M), lipid droplets (L), and numerous membranes of the endoplasmic reticulum (*) are evident. The surface of these cells contains microvilli which project into pericapillary spaces (Ps). The endothelial wall is provided with numerous microvilli and other cellular projections (arrows). Capillary lumen = Ca. (x 4,600; rat, mid-pregnancy).

C Transmission electron micrograph of an intercellular canaliculus (Ca) limited by a number of well-differentiated luteal cells. These luteal cells display microvilli (Mv) which project into the canal (arrow). Note both the rather close relationship between mitochondria (M) and a lipid droplet (L) and the presence of small granules (arrow *) that likely represent (micro)peroxisomes. Tight and gap junctions are also evident between adjacent luteal cells (double arrow). N = nucleus. (x 46,800; rat, midpregnancy).

D Transmission electron micrograph of whorls of smooth-surfaced endoplasmic reticulum in a mature luteal cell. (x 2,550; rabbit, midpregnancy).

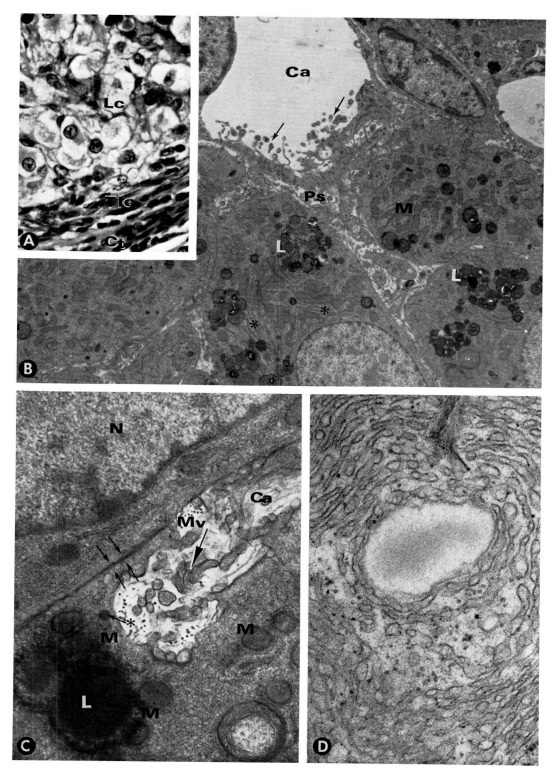

Plate 36. The Corpus Luteum.

Plate 37. Luteal Cells.

A   High-magnification SEM of a luteal cell during midpregnancy in the rat. The prominent feature of the luteal cell surface is the presence of numerous spheroidal blebs (arrows) and corresponding and complementary deep invaginations (*). The invaginations represent areas of the cell surface that previously had been occupied by the projections of an apposing cell. (x 7,400).

B   This stereo view of the surface of luteal cells in a mouse corpus luteum on day 4 of pregnancy depicts the rather thick, villous-like projections (arrows) that in the living state infold deeply into the cytoplasm of apposing luteal cells to form the characteristic invaginations presented in A. (x 5,600).

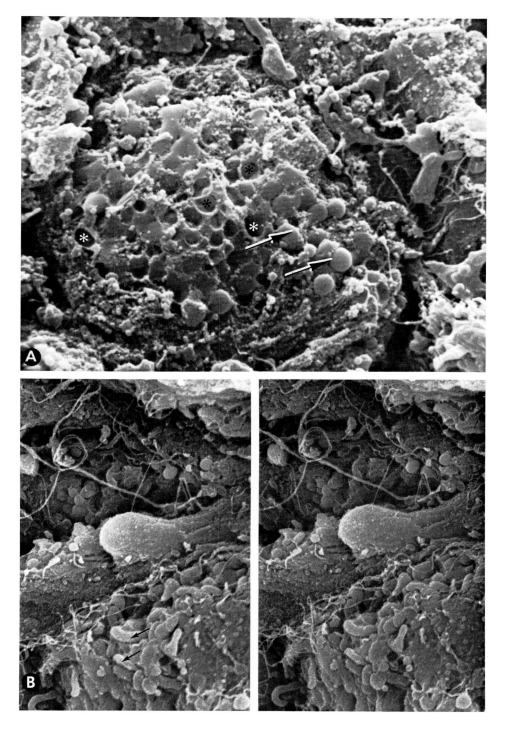

Plate 37. Luteal Cells.

Plate 38. The Corpus Albicans.

A   Low-magnification scanning electron micrograph of a corpus luteum (Cl) that has nearly completed transformation into a corpus albicans. Abundant connective tissue (Ct) is present within the luteal mass. In addition, a thick layer of connective tissue separates the luteal body from the surface of the ovary (Se). (x 40; rabbit at end of pregnancy).

B   As observed by scanning electron microscopy, the luteal mass during regression is a honeycomb-like structure in which each space contains a highly vacuolarized, degenerating cell (Lc). Connective tissue = Ct, large vessel = V. (x 504; rabbit at end of pregnancy).

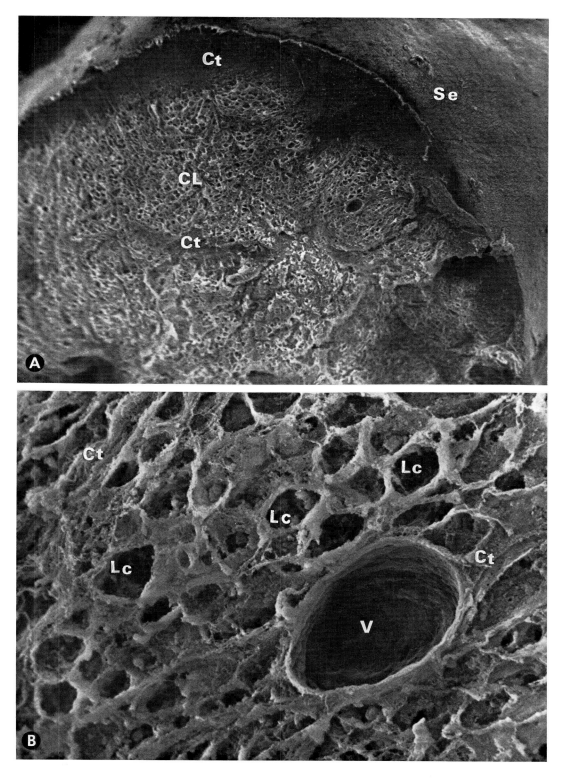

Plate 38. The Corpus Albicans.

3 The Related Ducts

3.1. The Fallopian Tubes

The Fallopian tubes, or oviducts, are paired muscular ducts that extend from the immediate vicinity of the ovary and communicate directly with the uterus. Depending upon the species, the oviducts vary considerably in length and degree of convolution. Embryonically, the Fallopian tubes arise from the cranial region of the primitive Müllerian ducts. During adult life, the oviducts provide (1) the channel through which spermatozoa and oocytes are transported, (2) the site of egg fertilization, and (3) during the early preimplantation stages of pregnancy, the essential physical and nutritional milieu for the cleaving embryo. For a more detailed discussion of the cellular, immunological, endocrinological, and biochemical properties of the mammalian oviduct, the reader's attention is directed to *Hafez* and *Blandau* (1969).

For the purpose of description, the oviduct is traditionally divided into four anatomical sections which, from the ovarian end, are (1) the preampulla (clearly evident only in some mammals), (2) the ampulla, (3) the isthmus, and (4) the junctura. The preampulla contains the ostium abdominale, the fimbriae, and the infundibulum. The most important adaptation of this region are the fimbriae. These finger-like projections form a funnel-like structure which more or less completely envelopes the ovary. In some species, the ovary is enclosed by a sac or bursa (bursa ovarii) consisting of a thin peritoneal fold of the mesosalpinx, which is attached to the upper oviduct. Generally, the wall of the preampulla is quite thin, and the cells of the epithelium are almost entirely ciliated. The function of this region is devoted to the transport of ovulated oocytes from the ovary to the site of fertilization, the ampulla.

The wall of the ampulla is somewhat thicker than the preampullary section (but only in mammals that have a clear preampulla, such as the pig), and its epithelium contains a comparatively reduced population of ciliated cells. The isthmus comprises the major portion of the oviduct (about 2/3) and has a pronounced muscular coat and an epithelium composed of primarily nonciliated cells. This region's major function is the transport of spermatozoa and fertilized

eggs. The junctura (intramural or interstitial section) is the region of the oviduct that crosses directly into the uterine cavity at the ostium uterinum tubae (uterotubal junction). The muscular layers of the junctura are well developed, and the epithelium contains a larger number of ciliated cells than encountered in the isthmus. The lumen of the junctura is the narrowest of the entire oviduct.

The oviductal wall is composed of an external serosa (tunica serosa), a middle muscular layer (tunica muscularis), and an internal mucosa (tunica mucosa) (Plate 39 A). The mucosa of the ampulla is deeply evaginated and folded into a complex, ramified system of plicae which, in cross section, form a labyrinthine network. By contrast, the mucosa of the isthmus is not extensively infolded and contains simple, longitudinal folds (Plate 40 A). The mucosal membrane is a pseudostratified epithelium composed of a layer of columnar cells attached to a basal lamina which is in turn attached to a richly vascular connective tissue. Two types of cells are present in this epithelium: ciliated and nonciliated. There is no discernible regularity in the distribution of these cells, although, as previously mentioned, ciliated cells are more abundant in the preampullary and ampullary regions than in the isthmus or junctura. In areas of the oviduct populated by ciliated cells, these elements typically occupy apical and lateral positions on the mucosal folds. By contrast, nonciliated cells are generally localized in the grooves or basal portions of the folds (Plate 39 B).

The cytoplasm of ciliated cells contains numerous ribosomes; cisternae of rough-surfaced endoplasmic reticulum; elongated, electron-dense mitochondria; and Golgi complexes localized in the apical regions of the cells. The characteristic feature of these cells is the abundant cilia that display the usual 9 + 2 microtubular arrangement and regularly oriented basal body complexes (*Brower* and *Anderson,* 1969; *Nilsson* and *Reinius,* 1969; *Dirksen* and *Satir,* 1972) (Plate 39 C). Typically, a number of short microvilli are present among the cilia (Plate 40); by transmission electron microscopy they are observed to be covered with a "fuzzy coat" or glycocalyx (Plates 39 D, 41 B).

Ostial and ampullary cilia are commonly believed

109

to beat in the direction of the uterus, and a general impression of their undulating nature is most apparent by scanning electron microscopy (*Ferenczy* and *Richart,* 1974) (Plates 40 B, 41). As kinetic organelles (*Ferenczy,* 1974; *Hafez,* 1977), the beating action of cilia in the fimbriae has a central role in the transport of the ovulated oocyte(s) from the surface of the ovary to the ampulla, and, in other regions, they facilitate the passage of the fertilized ovum through the oviduct and into the uterus. In several mammals, however, not all oviductal cilia beat in the direction of the uterus (*Gaddum-Rosse* and *Blandau,* 1976). Some groups of ciliated cells, especially those present on the longitudinal folds of the mucosa in the isthmus, beat in the direction of the ovary, whereas other cilia, present on the surface of adjacent ridges, beat toward the uterus. *Blandau* (1973) has suggested that alternate ciliary movements function in the transport of spermatozoa toward the ovary and in the passage of the ovum toward the uterus. In addition to ciliary action, peristaltic contractions of the tunica muscularis also contribute to the movement of gametes and preimplantation embryos (*Talo,* 1974). Finally, both the maintenance and full development of oviductal cilia are strictly dependent on the continuous presence of steroid hormones, since ovariectomy results in the nearly complete deciliation of the oviductal epithelium (*Brenner,* 1969; *Ferenczy, Richart, Agate, Purkerson,* and *Dempsey,* 1972; *Rumery* and *Eddy,* 1974; *Verhage* and *Brenner,* 1975).

The nonciliated cells of the oviductal epithelium are primarily secretory elements. By transmission electron microscopy, they appear as columnar cells containing numerous free ribosomes, cisternae of the rough-surfaced endoplasmic reticulum, and a prominent Golgi complex which is always located in a supranuclear position and directed toward the lumen of the oviduct (*Hashimoto, Shimoyama, Mori, Komori, Tomita,* and *Akashi,* 1959; *Nilsson* and *Rutberg,* 1960). During each reproductive cycle, the cytoplasm of these cells undergoes major fine structural alterations. In a comprehensive transmission electron microscopic study of the rabbit oviduct during the pre- and postcoital phases of the reproductive cycle, *Brower* and *Anderson* (1969) observed that numerous electron-dense granules accumulated in the nonciliated cells during the precoital stages. The formation of these granules involved ribosomes, elements of rough-surfaced endoplasmic reticulum, and, finally, packaging in the Golgi complex. Following coitus, when ova and early cleaving embryos passed down the oviduct, many nonciliated cells assumed a goblet shape, with the apical portions of the cells packed with granules. These granules were observed either isolated within the cytoplasm or in the process of coalescing and/or of being discharged (with an encompassing membrane) into the lumen (Plate 39 A). After the discharge of the granules has been completed, the secretory cells assume an elongated form but persist and take part in the ensuing reproductive cycle.

As observed by scanning electron microscopy, the secretory cells of the oviduct have a spherical or elliptical crown populated by microvilli that vary considerably in number (*Ferenczy* and *Richart,* 1974) (Plates 40 B, 41). When viewed from the cell surface, the secretion of granules in the postcoital oviduct involves the formation of small apical protrusions and/or larger bulbous structures accompanied by the partial or total disappearance of microvilli (Plates 40 B, 41). Histochemical studies demonstrate that the secretory product of the oviductal mucosa is mucoprotein in nature (*Balboni,* 1953; *Zachariae,* 1958; *Fredricsson,* 1969; *Brower* and *Anderson,* 1969). The function(s) of the secretions are, however, somewhat obscure, although it has been suggested that they may be required for the maintenance of (1) the fertilizable state of the gametes, (2) the epithelium itself, and (3) both the maintenance and protection of the preimplantation embryo (such as in the rabbit where the secretions are likely involved in the deposition of the mucin coat) (*Glass* and *McClure,* 1965; *Bishop,* 1969; *Greenwald,* 1969).

Finally, the musculature of the oviduct is composed of an inner circular and an external longitudinal layer. A third layer of longitudinal fibers is usually present in the region of the isthmus and especially near the uterus. The thickness of the muscular layers generally increases from the infundibulum toward the uterus. The entire oviduct is covered by a thin lamina serosa.

3.2 The Uterus

The uterus is a comparatively thin-walled, muscular organ that arises as a specialized portion of the Müllerian duct. Rostrally, the uterus communicates directly with the oviducts, and caudally, it opens into the vagina. The uterus may be divided into several anatomical sections:

1. the fundus or region of the uterus in some Edentates and Primates, including humans, which extends above the point of entry of the oviduct,
2. the expanded midportion or body of the uterus,

3. a lower, constricted portion or cervix which protrudes through and opens into the vagina by means of the cervical canal.

In cross section, the organization of the uterine wall is similar to that of the oviduct in that it is composed of three layers: a mucosa or endometrium, a muscularis or myometrium, and an external serosa or perimetrium. For a comprehensive discussion of the significant variations in the comparative embryology and anatomy of the mammalian uterus, the works of *Wynn* (1967, 1977) are suggested.

The major experimental emphasis for morphological, molecular, and physiological studies of the mammalian uterus has been devoted to the cyclical changes that take place in the endometrium during the reproductive cycle and pregnancy. Consequently, in this section, primary interest involves the morphophysiological alterations of the endometrium. The endometrium consists of a simple, columnar epithelium that is quite similar to that observed in the oviduct. The columnar cells are attached to a basal lamina that is relatively thick, but, in contrast to the oviductal mucosa, the subjacent lamina propria is quite extensive and contains numerous glands and a rich vascular network. Both ciliated and nonciliated (secretory) cells compose the endometrial epithelium (Plates 42, 43 A). In addition, numerous glands are randomly distributed throughout the lamina propria (Plate 42 A), and the orifices of the glands, surrounded by clusters of ciliated cells (Plate 43), open in an irregular fashion onto the mucosal surface. The glandular openings are most abundant in the upper regions of the uterus. Unlike the oviduct, the uterus contains a reduced population of ciliated cells. Generally, ciliated cells appear as isolated elements or are massed in small groups (Plate 43 A). Both their density and number are highly variable between different regions of the uterus.

In humans, randomly spaced groups of ciliated cells are present along the mucosal lining of the cervical canal, whereas in the rabbit, these areas are comparatively more densely populated with ciliated cells. Furthermore, the mucosa of the human external os of the cervical canal, as well as the portion of the cervix that extends into the vagina (portio vaginalis), is covered by a squamous, stratified epithelium similar to that present in the vagina. In the rabbit, however, these same areas contain groups of ciliated cells similar to those observed within the cervical canal (Plate 45 B).

Both the structure and function of the endometrium (including the epithelium, lamina propria, glands, and vessels) are subject to the influence of ovarian steroid hormones. Generally, three stages of cellular reorganization are observed by light and transmission electron microscopy during each estrous or menstrual cycle: (1) a proliferative phase, (2) a secretory phase, and (3) a menstrual phase. Extensive light and electron microscopic studies by *Davies* and *Hoffman* (1973, 1975) have indicated the existence of six definable phases of cellular reorganization of the progestational endometrium of the rabbit. These phases include:

1. a priming phase in which no obvious cytological changes are observed;
2. a proliferative phase in which the endometrium is mitotically active and the columnar epithelium becomes pseudostratified (this stage is correlated with rising levels of estrogen, 20 α-hydroxypregn-4-en-3-one and progesterone);
3. a phase of epithelial relayering and folding that corresponds to the secretory phase in which the epithelium again becomes simple columnar with maximal mucosal folding (the secretory phase is under the increasing dominance of estrogen);
4. a stage of cell fusion in which the columnar cells become multinucleated and which coincides with the full maturity of the corpora lutea and maximal progesterone synthesis;
5. a maximal progestational phase; and
6. a declining phase during which the endometrium returns to its resting or estrous state if fertilization has not occurred. This phase is associated with the regression of the corpora lutea, declining levels of progesterone, and rising levels of 20 α-hydroxypreg-4-en-3-one.

Similar electron microscopic observations have been summarized by *Lawn* (1974) in a comprehensive examination of the endometrium during the sexual cycles of the human, cow, rat, mouse, guinea pig, and rabbit.

With the application of the scanning electron microscope, alterations in the surface architecture of the epithelium during the course of the reproductive cycle are readily visualized. The superficial appearance of both ciliated and nonciliated cells is quite similar to that described for the oviduct. The apices of nonciliated cells contain numerous microvilli, and often a single, isolated cilium is observed (Plate 43 B). Frequently, delicate strands of an amorphous material (fuzzy coat or glycocalyx) are attached to the external borders of the microvilli. During the secretory phase of the reproductive cycle, and primarily in those cells bordering the lumen of the uterine glands (glandular cells), the apical surfaces of nonciliated cells undergo dramatic ultrastructural reorganizations. The microvilli present on these cells either partially or totally disappear, and an increasing number of small blebs and/or larger bulbous projections appear on the cell surface. The blebs are eventually discharged from the cell membrane and lie on the surface of the endometrium (Plate 44).

The changes in the cell surface observed by scanning electron microscopy correspond to secretory processes similar to, if not more pronounced than, the secretion of material by the nonciliated cells of the oviductal mucosa (*Brower* and *Anderson, 1969*). During the secretory phase, the apical portion of the glandular cells contains pronounced and dilated Golgi complexes (Plate 42 B). Mitochondria are rather enlarged, and in some species, such as in humans, patches of glycogen are evident. In the rabbit, cisternae of rough-surfaced endoplasmic reticulum are not abundant, although numerous ribosomes and polysomes are observed. Further, within the cytoplasm of the glandular cells, electron-dense vacuoles as well as "characteristic" mucoid storage granules are evident (Plate 42 C). The fine structure of (nonciliated) surface cells and glandular cells differs considerably during the secretory phase. These differences have been interpreted as indicating that the secretions discharged into the uterine lumen arise primarily from the gland cells (*Davies* and *Hoffman, 1975*). Although the mode of secretion by uterine gland cells has been a subject of controversy (apocrine vs. merocrine), it appears most likely that apocrine secretion is the primary means by which material is discharged onto the epithelium (*Wynn, Harris,* and *Wooley, 1967; Sengel* and *Stoebner, 1970; Lawn, 1974; Motta* and *Andrews, 1976*). Scanning electron microscopic studies reveal the presence of cells that protrude to a varying extent above the mucosa and into the lumen. Some of the cells likely represent old, degenerated, or damaged secretory and ciliated cells that are in the process of sloughing off from the epithelium (*Johannison* and *Nilsson, 1972; Motta* and *Andrews, 1976*).

Transmission and scanning electron microscopic observations demonstrate that some of the cells on the surface of the uterine lumen of different mammals develop apical protrusions either during the preimplantation stages of pregnancy or during delayed implantation (*Nilsson, 1966; Beier, Petry,* and *Kuhnel, 1970; Meyer, 1970; Psychoyos* and *Mandon, 1971; Nilsson, 1972; Bergstrom* and *Nilsson, 1973; Aitken, 1975; Motta* and *Andrews, 1976*). Several investigators have shown that the apical protrusions pinch off from the epithelial cells through an apocrine mechanism and have suggested that these secretions may provide "nourishment" for the embryo prior to implantation. In the rat (*Enders* and *Nelson, 1973; Parr* and *Parr, 1974*) and mouse (*Parr* and *Parr, 1977*), apical protrusions (termed "pinopods") of the uterine epithelium have been suggested as being involved in endocytosis rather than apocrine secretion. Although in some mammals, such as the rabbit, apical protrusions are often observed to contain secretory vesicles (Plate 42 E) and therefore may be related to secretory processes (*Meyer, 1970; Motta* and *Andrews, 1976*), it is not possible to exclude at present that some of these projections are actually involved in endocytotic processes (Plate 42 D) (*Davies* and *Hoffman, 1975; Barberini, Sartori, Van Blerkom,* and *Motta, 1978*).

The composition of the uterine secretions during the reproductive cycle varies between species, and except for certain laboratory animals and humans, little is known about their precise biochemical nature (*Daniel, 1971*). In general, however, the secretions that line the uterine mucosa during the progestational phase are rich in carbohydrates, mucopolysaccharides, and proteins (*Lawn, 1974*). In the mouse, rat, rabbit, and human, the protein component of the secretions changes both qualitatively and quantitatively during the progestational period, especially during the development of the blastocyst (*Daniel, 1971; Van Blerkom, Manes,* and *Daniel, 1973; Surani, 1975*). Recently, it has been reported that the ionic composition of the mouse uterus also undergoes marked alteration during the preimplantation stages of pregnancy (*Borland, Hazra, Biggers,* and *Lechene, 1977*). Although the physiological significance of the secretions is not completely understood, it appears likely that they have a central role in embryonic development prior to implantation (*Daniel, 1971; Van Blerkom* and *Manes, 1977*). Furthermore, in species in which capacitation of spermatozoa is a prerequisite for fertilization (Chapter 4), the uterine secretions may be involved in an alteration of the plasma membrane overlying the acrosome, which leads to the fertilizable state of the male gamete (*Motta* and *Van Blerkom, 1975*). Although the function of the ciliated cells of the uterus is not as apparent as that of the oviduct, it seems likely that these elements participate in the release and distribution of secretory material throughout the mucosa. This hypothesis is supported by scanning electron microscopic observations (Plate

44) (*Kanagawa* and *Hafez,* 1973; *Motta* and *Andrews,* 1976; *Hafez,* 1977).

The nonciliated cells of the mucosa that line the cervical canal actively synthesize and secrete mucus that changes both chemically and physically during the reproductive cycle (*Blandau,* 1973). The function(s) of these secretions is also not entirely clear. A comparatively extensive population of ciliated cells is present in the cervical canal, and the cilia beat in the direction of the vagina (*Kanagawa* and *Hafez,* 1973; *Hafez,* 1977). The maintenance of cervical cilia does not appear to require ovarian steroid hormones as was noted for the oviduct (*Riches, Rumery,* and *Eddy,* 1975). However, the synthesis and secretion of cervical mucus by nonciliated cells are under the influence of ovarian steroids since the production of mucus increases significantly during steroid stimulation (*El-Banna, Hafez,* 1972; *Riches, Rumery, Eddy,* 1975).

Finally, both the oviductal and uterine epithelia may be involved in the sequestration and phagocytosis of spermatozoa. Normally, spermatozoa, even a few hours following coitus, are observed to be phagocytized by leukocytes that invade the uterine mucosa at this time (*Austin,* 1957; *Yanagimachi* and *Chang,* 1963; *Zamboni,* 1971). However, the phagocytosis of spermatozoa has also been observed to involve the epithelium of the uterine glands and the oviductal mucosa (*Austin,* 1960; *Zamboni,* 1971; *Chakraborty* and *Nelson,* 1975). Whether such processes are important in the removal of nonviable, abnormal, or surplus spermatozoa from the genital tract remains unknown.

3.3 The Vagina

The vagina is a fibromuscular tube that embraces the neck of the uterus and connects directly with the external genitalia. The vaginal wall is composed of a mucosa, a muscularis, and a layer of dense connective tissue. The mucosa is evaginated into a number of transverse folds called rugae and consists of a stratified squamous epithelium. The cells of the vaginal epithelium are flattened and accumulate mucus and glycogen during the secretory phase of the reproductive cycle. In the rabbit, groups of ciliated cells are present in the upper regions of the vagina (Plate 45 B), where it communicates with the cervix and forms a deeply infolded region, the fornix vaginae (Plate 45 A). In the lower portions of the vagina, epithelial cells are polyhedral and contain numerous, prominent microvilli (Plate 46). Primarily during the phase of maximum progesterone synthesis, the apical surface of numerous epithelial cells contains ruffles, bulbous evaginations, and smaller membranous blebs (Plate 46 B). It is likely that these evaginations are a consequence of the significant accumulation of mucus droplets within these cells that occurs during the luteal phase of the reproductive cycle (*Eddy* and *Walker,* 1969), but other evaginations may be related to endocytotic processes, as appears to occur in the uterine and vaginal epithelia (phagocytosis of spermatozoa, for example) (*Correr, Motta* and *Hafez,* 1978).

In humans, the cyclic changes of the epithelium are not as obvious as in most laboratory animals, but in all mammals, the changes are strictly dependent upon the secretion of ovarian hormones. Under the influence of estrogen (follicular phase), the vaginal epithelium is usually quite keratinized and prone to slough off. During the luteal or progestational phase, mucus that accumulates in the apical portions of cells composing the vaginal epithelium is discharged into the vaginal lumen, resulting in the mucification of the mucosa (Plate 46 B). During this phase, the cells of the vaginal mucosa are cuboidal and are covered with microvilli to which a dense, filamentous material (mucus) is attached (Plate 46 B). Finally, the superficial and lateral aspects of the epithelial cells of the rat and mouse vagina, as observed by scanning electron microscopy, display complex and interconnected microridges (microrugae or microplicae; *Parakkal,* 1974; *Rubio,* 1976). Microplicae are observed in other stratified or partially keratinized epithelia and are thought to reduce friction between the mucosal surface and solid material (such as in the conjunctiva and esophagus) (*Parakkal,* 1974; *Motta, Andrews,* and *Porter,* 1977).

References

Aitken, R. J. (1975). Ultrastructure of the blastocyst and endometrium of the roe deer (Capreolus capreolus) during delayed implantation. J. Anat. (London) *119:*369–384.

Austin, C. R. (1957). Fate of spermatozoa in the uterus of the mouse and rat. J. Endocrinol. *14:*335–342.

Austin, C. R. (1960). Fate of spermatozoa in the female genital tract. J. Reprod. Fert. *1:*151–156.

Balboni, G. (1953). Ulteriori richerche istochimiche sull'epitelio tubarico dell donna. Boll. Soc. Ital. Biol. Sper. *29:*1394–1399.

Barberini, F., Sartori, S., Van Blerkom, J., and *Motta, P.* (1978). Changes in the surface morphology of the rabbit endometrium relates to the estrous and progestational stages of the reproductive cycle: A scanning and transmission electron microscopic study. Cell. Tiss. Res. *190:* 207–222.

Beier, H. M., Petry, G., and *Kuhnel, W.* (1970). Endometrial secretion and early mammalian development. In: Mammalian Reproduction, pp. 264–285 (Gibian, H. and Plotz, E. J., eds.).

Bergstrom, S., and *Nilsson, O.* (1973). Various types of embryo-endometrial contacts during delay of implantation in the mouse. J. Reprod. Fert. *32:*531–533.

Bishop, D. W. (1969). Sperm physiology in relation to the oviduct. In: The Mammalian Oviduct. Comparative Biology and Methodology (Hafez, E. S. E., and Blandau, R. J., eds.). University of Chicago Press, Chicago, London.

Blandau, R. J. (1973). The female reproductive system. In: Histology, pp. 761–806 (Greep, R. O., and Weiss, L., eds.). McGraw-Hill, New York.

Borland, R. M., Hazra, S., Biggers, J. D., and *Lechene, C. P.* (1977). The elemental composition of the environment of the gametes and preimplantation embryo during the initiation of pregnancy. Biol. Reprod. *16:*147–157.

Brenner, R. M. (1969). The biology of oviductal cilia. In: The Mammalian Oviduct. Comparative Biology and Methodology (Hafez, E. S. E., and Blandau, R. J., eds.). University of Chicago Press, Chicago and London.

Brower, L. K., and *Anderson, E.* (1969). Cytological events associated with the secretory process in the rabbit oviduct. Biol. Reprod. *1:*130–148.

Chakraborty, J., and *Nelson, L.* (1975). Fate of surplus sperm in the Fallopian tube of the white mouse. Biol. Reprod. *12:*455–463.

Correr, S., Motta, P., and *Hafez, E. S. E.* (1978). Phagocytosis of spermatozoa in rabbit vaginal epithelium as revealed by transmission and scanning electron microscopy. Proc. 5th. Europ. Cong. Fert. Steril.

Daniel, J. C. (1971). Uterine proteins and embryonic development. In: Advances in the Biosciences, Vol. 6, pp. 191–206 (G. Raspe, ed.). Pergamon Press, New York.

Davies, J., and *Hoffman, L. H.* (1973). Studies on the progestational endometrium of the rabbit. I. Light microscopy, day 0 to day 13 of gonadotrophin-induced pseudopregnancy. Am J. Anat. *137:*423–445.

Davies, J., and Hoffman, L. H. (1975). Studies on the progestational endometrium of the rabbit. II. Electron microscopy, day 0 to day 13 of gonadotrophin-induced pseudopregnancy. Am. J. Anat. *142:*335–366.

Dirksen, E. R., and *Satir, P.* (1972). Ciliary activity in the mouse oviduct as studied by transmission and scanning electron microscopy. Tissue and Cell *4:*389–404.

Eddy, E. M., and *Walker, B. E.* (1969). Cytoplasmic fine structure during hormonally controlled differentiation in the vaginal epithelium. Anat. Rec. *164:*205–218.

El-Banna, A., and *Hafez, E. S. E.* (1972). The uterine cervix in mammals. Am J. Obstet. Gynec. *112:*145–164.

Enders, A. C., and *Nelson, D. M.* (1973). Pinocytotic activity of the uterus of the rat. Amer. J. Anat. *138:*277–300.

Ferenczy, A. (1974). The surface ultrastructure of the human Fallopian tube. A comparative morphophysiologic study. In: Scanning Electron Microscopy (O. Johari, ed.). I.I.T. Research Institute, Chicago.

Ferenczy, A., and *Richart, R.* (1974). Female Reproductive System: Dynamics of Scan and Transmission Electron Microscopy. John Wiley & Sons, New York.

Ferenczy, A., Richart, R., Agate, F., Purkerson, M., and *Dempsey, E.* (1972). Scanning electron microscopy of the human fallopian tube. Science *175:*783–784.

Fredricsson, B. (1969). Histochemistry of the oviduct. In: The Mammalian Oviduct. Comparative Biology and Methodology (E. S. E. Hafez and R. J. Blandau, eds.). University of Chicago Press, Chicago and London.

Gaddum-Rosse, P., and *Blandau, R. J.* (1976). Comparative observation on ciliary currents in mammalian oviducts. Biol. Reprod. *14:*605–609.

Glass, L. E., and *McClure, T. R.* (1965). Postnatal development of the mouse oviduct: Transfer of serum antigens to the tubal epithelium. In: The Preimplantation Stages of Pregnancy, Ciba Foundation Symposium (G. E. W. Wolstenholme and M. O'Connor, eds.). Little, Brown and Co., Boston.

Greenwald, G. S. (1969). Endocrinology of oviductal secretions. In: The Mammalian Oviduct. Comparative Biology and Methodology (E. S. E. Hafez and R. J. Blandau, eds.). University of Chicago Press, Chicago and London.

Hafez, E. S. E. (1977). Kinetics of luminal secretions in the female reproductive tract. Ultrastructural and physiological parameters. Acta Anat. (Basel) *97:*143–146.

Hafez, E. S. E., and *Blandau, R. J.* (1969). The Mammalian Oviduct. Comparative Biology and Methodology. University of Chicago Press, Chicago and London.

Hashimoto, M., Shimoyama, T., Mori, Y., Komori, A., Tomita, H., and *Akashi, K.* (1959). Electron microscopic observations on the secretory process in the Fallopian tube of the rabbit (report II). J. Japan Obstet. Gynecol. Soc. *6:*684–691.

Johannison, E., and *Nilsson, L.* (1972). Scanning electron microscopic study of the human endometrium. Fertil. Steril. *23:*613–625.

Kanagawa, H., and *Hafez, E. S. E.* (1973). Kinocilia and sperm dynamics in the cervix uteri of the rabbit. J. Reprod. Med. *10:*90–94.

Lawn, A. M. (1974). The ultrastructure of the endometrium during the sexual cycle. In: Advances in Reproductive Physiology Vol. 6 (H. W. H. Bishop, ed.). Elek Science, London.

Meyer, J. M. (1970). Recherches sur l'ultrastructure de la muqueuse uterine de la lapine. Arch. Anat. Histol. Embryol. Norm. Exper. *53:*1–40.

Motta, P., and *Andrews, P. M.* (1976). Scanning electron microscopy of the endometrium during the secretory phase. J. Anat. (London) *122:*315–322.

Motta, P., and *Van Blerkom, J.* (1975). A scanning electron microscopic study of rabbit spermatozoa in the female reproductive tract following coitus. Cell. Tiss. Res. *163:*29–44.

Motta, P., Andrews, P. M., and *Porter, K. R.* (1977). Microanatomy of Cells and Tissue Surfaces: An Atlas of Scanning Electron Microscopy. Vallardi and Lea Febiger, Milano and Philadelphia.

Nilsson, O. (1966). Structural differentiation of luminal membrane in rat uterus during normal and experimental implantations. Z. Anat. Entwicklungsgesch. *125:*152–159.

Nilsson, O. (1972). Ultrastructure of the process of secretion in the rat uterine epithelium at preimplantation. J. Ultrastr. Res. *40*:572–580.

Nilsson, O., and Reinius, S. (1969). Light and electron microscopic structure of the oviduct. In: The Mammalian Oviduct. Comparative Biology and Methodology (E. S. E. Hafez and R. J. Blandau, eds.). University of Chicago Press, Chicago and London.

Nilsson. O., and Rutberg, U. (1960). Ultrastructure of secretory granules in postovulatory rabbit oviduct. Exp. Cell Res. *21*:622–625.

Parakkal, P. F. (1974). Cyclical changes in the vaginal epithelium of the rat seen by scanning electron microscopy. Anat. Rec. *178*:529–538.

Parr, M. B., and Parr, E. L. (1974), Uterine luminal epithelium: Protrusions mediate endocytosis not apocrine secretion in the rat. Biol. Reprod. *11*:220–233.

Parr, M. B., and Parr, E. L. (1977). Endocytosis in the uterine epithelium of the mouse. J. Reprod. Fert. *50*:151–153.

Psychoyos, A., and Mandon, P. (1971). Scanning electron microscopy of the surface of the rat uterine epithelium during delayed implantation. J. Reprod. Fert. *26*:137–138.

Riches, W. G., Rumery, R. E., and Eddy, E. M. (1975). Scanning electron microscopy of rabbit cervix epithelium. Biol. Reprod. *12*:573–583.

Rubio, C. A. (1976). The exfoliating cervico-vaginal surface. II. Scanning electron microscopical studies during the estrous cycle in mice. Anat. Rec. *185*:359–372.

Rumery, R. E., and Eddy, E. M. (1974). Scanning electron microscopy of the fimbriae and ampullae of rabbit oviducts. Anat. Rec. *178*:83–102.

Sengal, A., and Stoebner, P. (1970). Ultrastructure de l'endometre humain normal. II. Les glandes. Z. Zellforsch. mikrosk. Anat. *109*:260–277.

Surani, M. A. H. (1975). Hormonal regulation of proteins in the uterine secretion of ovariectomized rats and the implications for implantation and embryonic diapause. J. Reprod. Fert. *43*:411–417.

Talo, A. (1974). Electric and mechanical activity of the rabbit oviduct *in vitro* before and after ovulation. Biol. Reprod. *11*:335–345.

Van Blerkom, J., and Manes, C. (1977). The molecular biology of preimplantation embryogenesis. In: Concepts in Early Mammalian Development (M. I. Sherman, ed.). M.I.T. Press, Cambridge, Mass.

Van Blerkom, J., Manes, C., and Daniel, J. (1973). Development of preimplantation rabbit embryos *in vivo* and *in vitro*. I. An ultrastructural comparison. Devel. Biol. *35*:262–282.

Verhage, H. G., and Brenner, R. M. (1975). Estradiol-induced differentiation of the oviductal epithelium in ovariectomized cats. Biol. Reprod. *13*:104–111.

Wynn, R. M. (1967). Cellular Biology of the Uterus. Appleton-Century-Crofts, New York.

Wynn, R. M. (1977). Biology of the Uterus. Plenum Press, New York and London.

Wynn, R. M., Harris, J. A., and Wooley, R. S. (1967). Ultrastructural cyclic changes in the human endometrium. II. Normal postovulatory phase. Fertil. Steril. *18*:721–738.

Yanagimachi, R., and Chang, M. C. (1963). Infiltration of leucocytes into the uterine lumen of the golden hamster during the oestrous cycle and following mating. J. Reprod. Fert. *5*:389–396.

Zachariae, F. (1958). Autoradiographic (^{35}S) and histochemical studies of sulphomucopolysaccharides in the rabbit uterus, oviducts and vagina. Acta Endocr. *29*:118.

Zamboni, L. (1971). Fine Morphology of Mammalian Fertilization. Harper and Row, New York and London.

Plate 39. Oviductal Mucosa.

A   Transmission electron micrograph of the epithelium and tunica muscularis (Tm) of a mouse oviduct during the secretory phase of the reproductive cycle. This section illustrates the appearance of secretory cells in the mucosa as well as the presence of presumed secretory granules (or possibly lipids) within the cytoplasm and on the surface of the cells (Sg). (x 1,000).

B   Transmission electron micrograph of nonciliated cells in the oviductal epithelium during the secretory phase of the cycle. Numerous microvilli (Mv), elongated mitochondria (M), and an extensive Golgi complex (G) are evident. Nucleus = N. (x 1,200; mouse).

C   High-magnification transmission electron micrograph of oviductal cilia demonstrating the characteristic 9 + 2 arrangement of microtubules. (x 10,000; mouse).

D   High-magnification transmission electron micrograph of microvilli present on the surface of secretory cells in a mouse oviduct during the secretory phase. The "fuzzy coat" of filamentous material likely corresponds to a glycocalyx. (x 57,000).

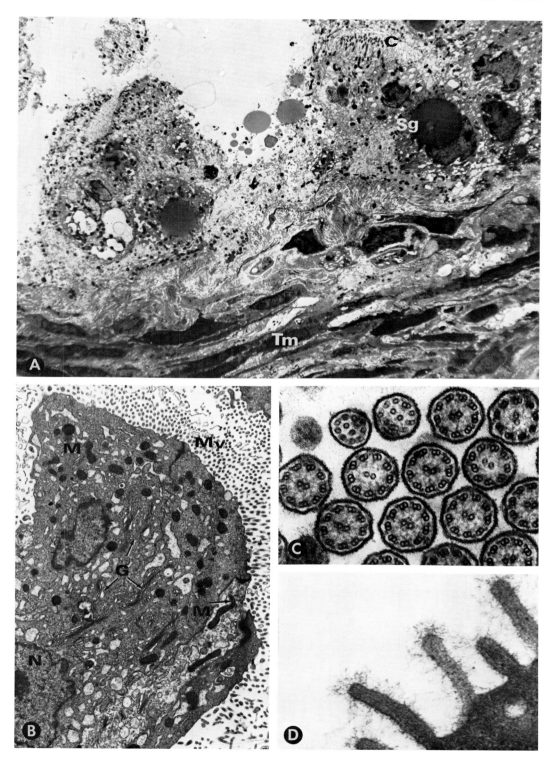

Plate 39. Oviductal Mucosa.

Plate 40. General View of Oviductal Mucosa.

A In this relatively low-magnification scanning electron micrograph of the isthmic region of the oviduct, both ciliated (Cc) and nonciliated secretory cells (Sc) are evident. The nonciliated cells appear to protrude into the lumen to a greater extent than do adjacent ciliated cells. (x 2,080; rabbit, 15 hours postcoitum).

B The overall appearance of ciliated (Cc) and nonciliated (Sc) cells is more apparent at higher magnification. Numerous microvilli are present on the surface of the nonciliated cells. (x 5,800; isthmic region of rabbit oviduct, 15 hours postcoitum).

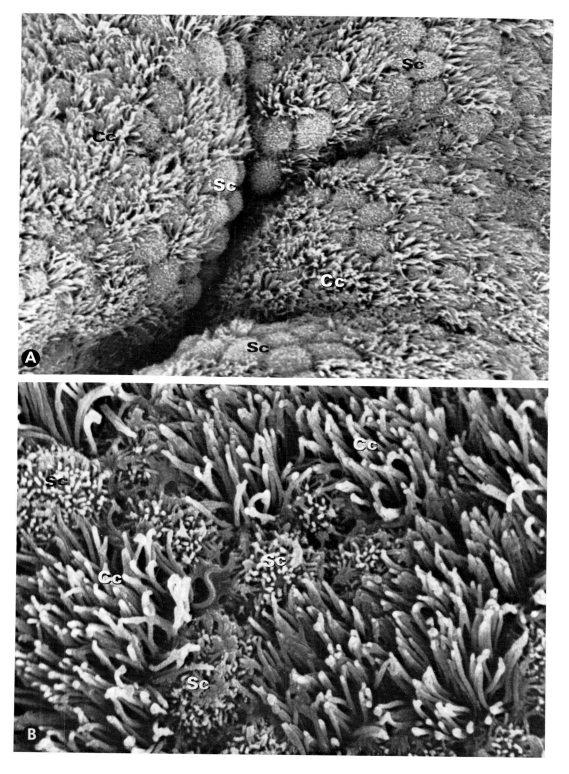

Plate 40. General View of Oviductal Mucosa.

Plate 41. Oviductal Mucosa (Secretory Phase).

A This scanning electron micrograph illustrates the appearance of secretory cells (Sc) in the rabbit oviduct at 36 hours postcoitum. Major features of interest are the numerous blebs and droplets (arrows) which likely represent secretory material. These cytoplasmic blebs are located on the surface of nonciliated cells and, when detached from the cytoplasm, are observed adhering to adjacent ciliated cells (arrow*). It is probable that the distribution of this material throughout the oviductal epithelium is facilitated by the cilia. (x 5,130).

B At higher magnification, the surface of nonciliated cells displays microvilli (Mv), blebs, and evaginations that arise from the apical surface (arrows) and are likely related to the secretory process. A thin, filamentous material is often observed attached to some microvilli (arrows*) and may represent a residual glycocalyx. (x 20,000; rabbit oviduct, 36 hours postcoitum).

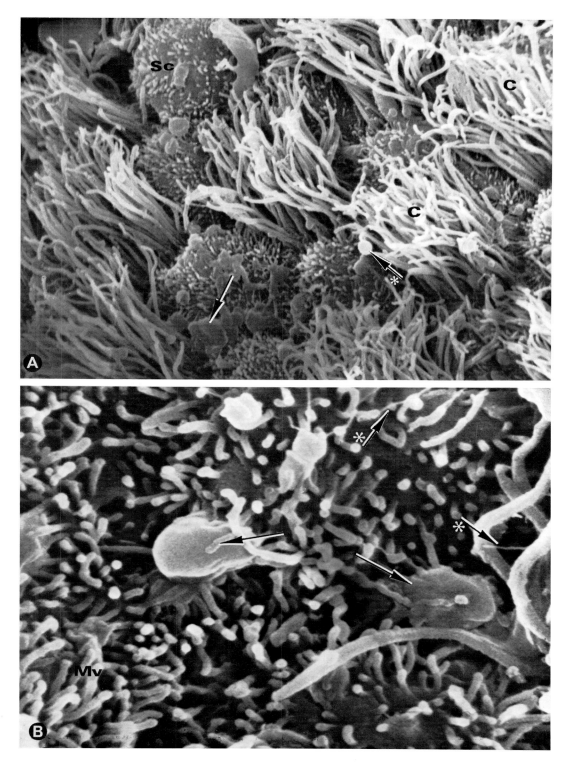

Plate 41. Oviductal Mucosa (Secretory Phase).

Plate 42. Uterus during the Secretory Phase.

A   This transmission electron micrograph illustrates the characteristic appearance and organization of a uterine gland during the secretory phase of the reproductive cycle. (x 1,800; mouse).

B   Nonciliated, glandular cells of the mouse uterus during the secretory phase are observed to contain numerous electron-dense droplets of varying size. These droplets (Sg) likely represent secretory material that is discharged during this phase of the reproductive cycle. The secretory process is more evident when observed by scanning electron microscopy (see Plate 44). (x 9,200).

C   Apical protrusion of a luminal cell probably related to a process of endocytosis. (x 11,000; rabbit, 24 hours postcoitum).

D   A typical feature of the glandular cells of the uterine epithelium is a rather well-developed Golgi complex (G) located at the apical region of the cytoplasm. Mitochondria = M; nucleus = N. (3,200; mouse uterus during secretory phase).

E   Large cellular projection of a uterine epithelial cell containing numerous secretory vesicles. (x 13,000; rabbit, 24 hours postcoitum).

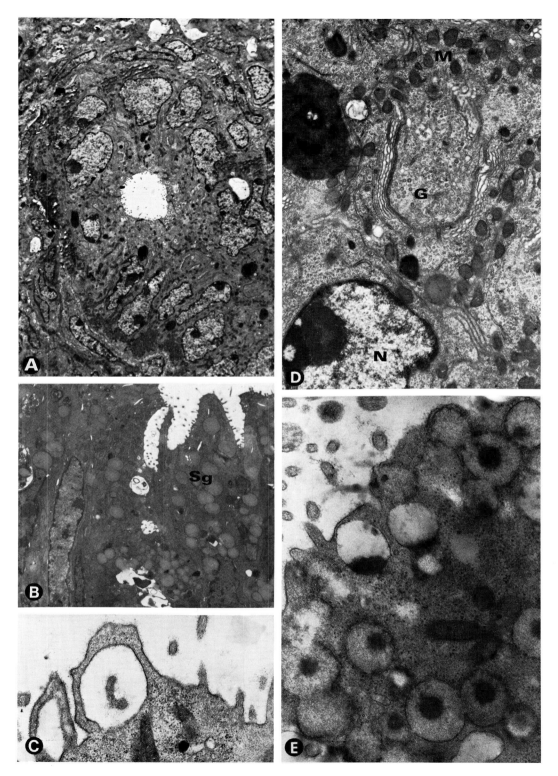

Plate 42. Uterus during the Secretory Phase.

Plate 43. The Endometrium.

A   Relatively low-magnification scanning electron micrograph of the endometrium in which ciliated cells (Cc), nonciliated cells (Sc), and the orifices of uterine glands are evident (arrows). (x 1,410; rabbit, 24 hours postcoitum).

B   Higher magnification view of a glandular opening bordered by microvillous cells (Mv). Occasionally, one or two short cilia (arrows) are observed protruding above surrounding microvilli. (x 3,000; rabbit, 24 hours postcoitum).

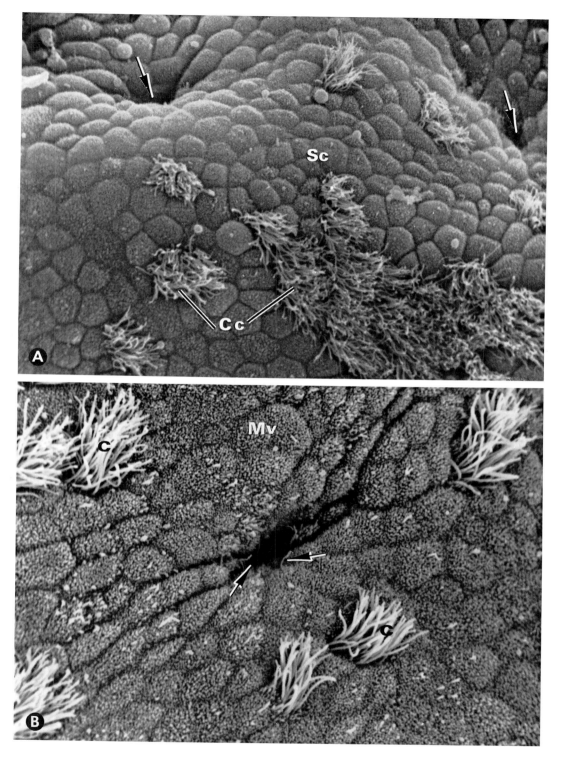

Plate 43. The Endometrium.

Plate 44. Endometrium (Secretory Phase).

A   The mucopolysaccharide/mucoprotein-containing secretion from the glandular cells of the endometrium is shown in this scanning electron micrograph. A large number of secretory cells surrounding a glandular opening are observed to contain blebs and bulbous projections on their apical surfaces as well as droplets that are not directly attached to the plasma membrane. These blebs and droplets (B) likely contain material that is stored by the gland cells, as is evident by transmission electron microscopy (Plate 42 B). Ciliated cells (Cc) are distributed about the orifice of the gland. (x 2,580; rabbit, 24 hours postcoitum).

B   Numerous bulbous projections are present on the surface of secretory cells lining the uterine epithelium during the luteal phase of the reproductive cycle. Some of these projections are apparently discharged from the cell membrane at the apical region of the cytoplasm by an apocrine-like mechanism (arrows). Note the absence of microvilli in a group of secretory cells in which a comparatively large bleb (B) is in the process of being discharged. Other protrusions (arrows*) possess an indented and ruffled surface and likely represent structures related to a process of endocytosis. (x 3,300; rabbit, 36 hours postcoitum).

126

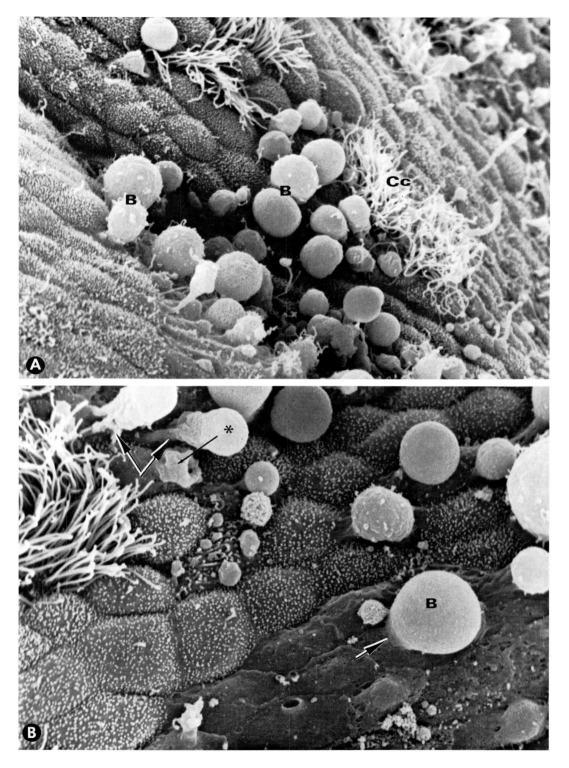

Plate 44. Endometrium (Secretory Phase).

Plate 45. Ectocervix and Fornix Vaginae.

A   A low-magnification view of the external os of the cervical canal (arrow) is illustrated in this scanning electron micrograph. Note the "flower-like" appearance of this area as a result of the presence of regularly arranged folds (*). (x 70; rabbit, 15 hours postcoitum).

B   The upper region of the vagina (fornix vaginae) is depicted in this low-magnification scanning electron micrograph. Both ciliated (Cc) and nonciliated (unmarked) cells are observed. The nonciliated cells may still represent secretory elements. In the rabbit, this region represents a zone of transition between the uterus and vagina in regard to the composition of the epithelium. The appearance of the fornix vaginae (distribution of ciliated and nonciliated cells) is much more closely related to that of the epithelial lining of the cervical canal than to the mucosa of the lower regions of the vagina. (x 1,400; rabbit, 36 hours postcoitum).

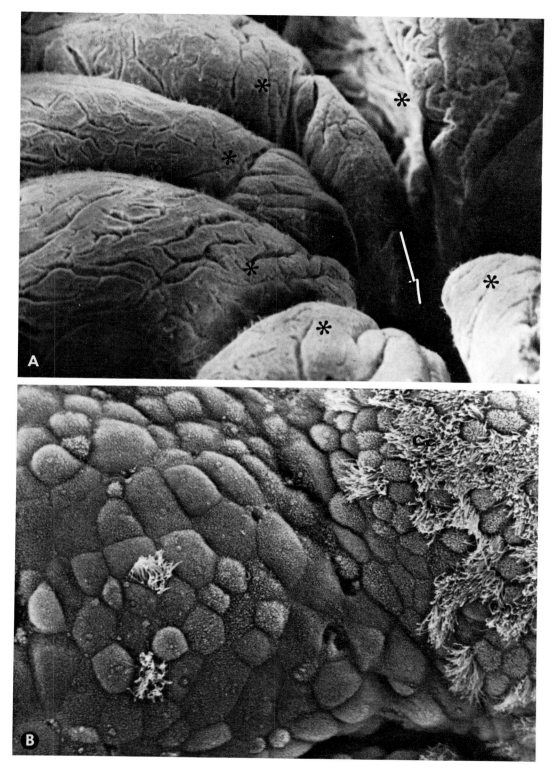

Plate 45. Ectocervix and Fornix Vaginae.

Plate 46. Vaginal Mucosa.

A The cells of the vaginal mucosa shown in this scanning electron micrograph are irregularly polyhedral and display the superficial appearance of a stratified epithelium. (x 850; rabbit, 15 hours postcoitum).

B At higher magnification, the cells of the vaginal epithelium are observed to contain numerous, tall microvilli (Mv) as well as irregular evaginations and cytoplasmic blebs (arrows). Some of these apical protrusions may contain mucus which may be in the process of being released from the cell membrane, thereby contributing to the mucification of the epithelium. Other protrusions with irregularly indented surfaces may be involved in a process of sequestration (endocytosis) as well. (x 7,500; rabbit, 24 hours postcoitum).

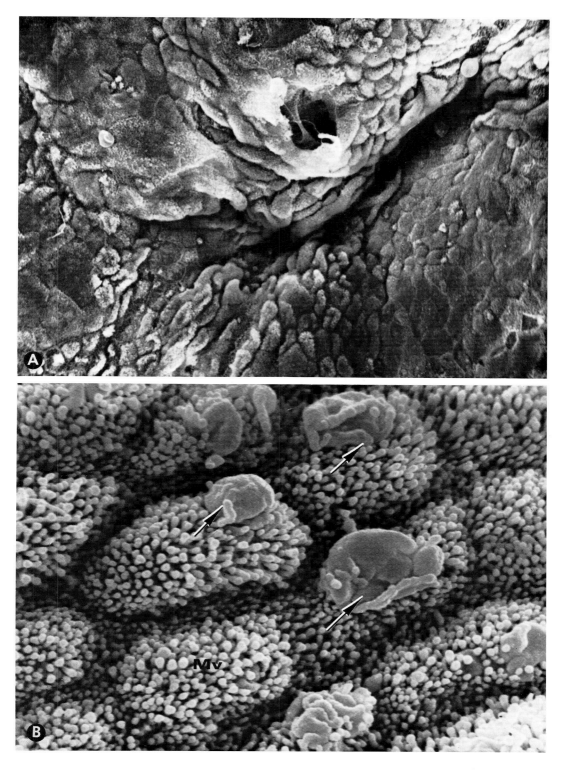

Plate 46. Vaginal Mucosa.

Section II

The Gametes and the Preimplantation Stages of Pregnancy

The following two chapters describe morphophysiological and morphogenetic processes involved in (1) the attainment of the fertilizable state of the gametes, and (2) the preimplantation stages of pregnancy. Although the observations presented in these chapters are derived from scanning and transmission electron microscopic studies, we have attempted to provide a biochemical and molecular biological background upon which these processes may be viewed. The application of elegant electron microscopic and biochemical techniques to the study of reproduction has considerably refined our knowledge of cellular and molecular aspects of early mammalian embryogenesis; however, as will become apparent in Chapter 5, our understanding of the developmental and functional significance of certain cytoplasmic structures is still unknown. Therefore, it is the purpose of these chapters not only to describe morphological processes involving gametes and preimplantation embryos, but also to indicate areas of cell structure and function relevant to early development in which our knowledge and understanding are incomplete.

4 Physiological and Morphological Aspects of Spermatozoa Leading to the Attainment of the Capacity to Fertilize an Oocyte

4.1. General Considerations

A spermatozoon is a highly motile cell whose structural organization and macromolecular composition effect passage through the cellular and noncellular layers of an ovulated oocyte ultimately to fuse with its plasma membrane. The function of spermatozoa is the transmission of the haploid, paternal genome to the oocyte. The gross morphology of a mature spermatozoon is highly species-specific, and an appreciation of the range of morphological diversity that exists may be obtained from the works of *Baccetti* (1970), *Phillips* (1974), and *Baccetti* and *Afzelius* (1976). Since it is the intent of this chapter to discuss morphophysiological events that occur on the surface of mammalian spermatozoa during postcoital residence in the female reproductive tract, the remarkable morphogenetic processes associated with the formation and maturation of spermatozoa during spermiogenesis are not discussed. For a comprehensive treatment of this subject, the excellent reviews of *Fawcett* (1970), *Zamboni* (1971), and *Phillips* (1974) are suggested.

Major reorganizations in both the composition and distribution of macromolecules take place within the membranous coverings of the mammalian spermatozoon during spermiogenesis. These changes are especially evident in the head region of a spermatozoon and have been documented by biochemical, immunological, and electron microscopic techniques. Some of the more prominent reorganizations include changes in the distribution and nature of proteins, glycoproteins (*Gordon, Dandekar,* and *Bartoszewicz,* 1975; *Millette,* 1977), antigenic sites (*Koehler* and *Perkins,* 1974; *Johnson,* 1975; *Koehler,* 1975; *Johnson* and *Howe,* 1975; *Koo, Boyse,* and *Wachtel,* 1977; *Millette* and *Bellve,* 1977), and clusters and ordered arrays of membranous particles (as observed by freeze-etch microscopy) (*Friend* and *Fawcett,* 1974; *Stackpole* and *Devorkin,* 1974; *Friend,* 1977). Since mammalian spermatozoa, after ejaculation, must penetrate through the cumulus mass (corona radiata, and the zona pellucida) in order to fuse with the vitelline membrane (*Austin,* 1975), it is not unlikely that pre-ejaculatory modifications may be required in order to convey a specific (species?) molecular and structural organization to spermatozoa, thus permitting these processes to be accomplished in an ordered sequence (*Johnson,* 1975; *Johnson* and *Howe,* 1975). Furthermore, in some mammals, structural and macromolecular alterations may continue in response to conditions and/or factors present in the female reproductive tract (*McRorie* and *Williams,* 1974; *Austin,* 1975; *Hartree,* 1977). Thus, many postcoital processes involving spermatozoa may be not only a consequence of, but also a continuation of, complex structural and biochemical events that began during spermiogenesis.

4.2. General Morphology

Most of our knowledge concerning the fine structural organization of mammalian spermatozoa has been derived from transmission electron microscopy (for example, *Fawcett,* 1970; *Phillips,* 1974). From an examination of the spermatozoa of numerous mammals, it is apparent that architectural similarities permit some basic generalizations to be made. Plate 47 demonstrates the anatomy of a generalized mammalian spermatozoon, as shown in diagrammatic form (Plate 47 A) and as observed in the mouse by TEM (Plate 47 B). The "typical" mammalian spermatozoon is composed of two gross morphological sections: the head and the tail. These sections may be further divided into the acrosomal and postacrosomal regions of the head, and the neck (connecting piece), middle (Plate 47 C), principal, and end pieces of the tail. Each of these regions is characterized by specific membranous organizations, organelles, intramembranous particles and macromolecules (*Friend,* 1977), and biochemical compositions (*McRorie* and *Williams,* 1974; *Johnson,* 1975; *Johnson* and *Howe,* 1975; *Morton,* 1977). Since a detailed, fine structural analysis of the entire spermatozoon is beyond the scope of this chapter, the reviews of *Fawcett* (1970) and *Phillips* (1974) are recommended.

4.3. Capacitation

A phenomenon of primary importance in the discussion of morphophysiological events occurring on the surface of mammalian spermatozoa during postcoital residence in the female reproductive tract, and one that is fundamental to the attainment of the fertilizable state of the male gamete, is capacitation. The observation that the spermatozoa of certain mammals require a period of time in the female reproductive tract before becoming capable of penetrating the cellular and noncellular investments of an ovulated oocyte was made independently by *Austin* (1951) and *Chang* (1951). The phenomenon was later termed "capacitation" by *Austin* (1952). Since the concept of capacitation was introduced, the process has been defined as a conditioning of spermatozoa within the female tract, which, when completed, permits the acrosome reaction (see section 4.4) to occur, usually in the immediate vicinity of the oocyte or within the cumulus cell mass. Numerous studies have indicated that a phenomenon similar to capacitation occurs *in vitro* if spermatozoa (of some species) are incubated in a suitable medium for a sufficient period of time (*Yanagamachi*, 1969a, 1969b, 1970; *Barros* and *Garavango*, 1970; *Gwatkin* and *Hutchison*, 1971; *Barros, Berrios*, and *Herrera*, 1973; *Bavister*, 1973; *Oliphant* and *Brackett*, 1973; *Brackett* and *Oliphant*, 1975). However, it is not entirely clear at present whether *in vivo* and *in vitro* capacitation are identical physiological processes. Finally, the universality of capacitation, defined as a period of conditioning in the female reproductive tract, is a controversial issue. Although capacitation has been shown to be a prerequisite for the fertilization of rabbit, rat, hamster, ferret, sheep, and pig oocytes (*Austin*, 1969), it may not be necessary for the fertilization of mouse, cow, monkey, and human oocytes (*Braden* and *Austin*, 1954; *Austin*, 1969; *Bryan*, 1974).

Biochemical and histological analyses indicate that during capacitation, both a glycoprotein coat and seminal plasma proteins may be removed from the acrosomal region of spermatozoa. These surface elements are most probably deposited on spermatozoa in the epididymis (*Bedford*, 1970; *Gordon, Dandekar*, and *Bartoszewicz*, 1975; *Hartree*, 1977). It has been suggested that this surface coat stabilizes the plasma membrane that overlies the acrosome and that an important step in capacitation lies in the removal of this material (*Austin*, 1975). The removal of the coat is thought to involve lytic agents present both within spermatozoa and also within the reproductive tract of the female (*Bedford*, 1970; *Soupart*, 1972; *Williams*, 1972). An important physiological correlate of capacitation is a marked increase in respiratory activity and intensity of motility of spermatozoa (*Hartree*, 1977).

Since capacitation appears to involve biochemical changes in the region of the plasma membrane that overlies the acrosome, it has long been thought that concomitant morphological changes should be evident and thus provide a structural basis for this physiological process. However, detailed TEM examinations of spermatozoa undergoing capacitation, and recovered by flushing the female reproductive tract with a suitable medium, have failed to demonstrate any morphological correlates (*Bedford*, 1970; *Bernstein* and *Teichman*, 1972). Two factors that may have resulted in the failure to detect morphological correlates are (1) that the recovery of spermatozoa from the reproductive tract by flushing may prelude the detection of subtle morphological changes that could be related to capacitation, and (2) that only a very limited number of spermatozoa and only a small portion of the plasma membrane per thin section are available for analysis by TEM.

The application of scanning electron microscopy to the study of the surface of spermatozoa has several advantages over the more traditional TEM approach. For example, a large population of spermatozoa may be examined *in situ* simultaneously, i.e., within different regions of the reproductive tract and at specific times following mating. Further advantages inherent in the use of SEM are (1) subtle morphological changes that may occur are more likely to be preserved, and (2) the entire surface area of the head of a spermatozoon may be examined. Consequently, the changes that are observed by SEM to occur on the surface of spermatozoa while they are resident in the female reproductive tract (section 4.5) may be related to capacitation or to other physiological and morphological processes that take place prior to fertilization such as spermatozoa transport, selection, or the acrosome reaction.

4.4. The Acrosome Reaction

The enzymatic composition of the mammalian acrosome indicates that this structure is a special type of secretory granule or a modified lysosome (*Allison* and *Hartree*, 1970; *Bernstein* and *Teichman*, 1972; *Hartree*, 1975). TEM studies of the origin and formation of the acrosome further strengthen this concept, because the acrosome originates from secretory granules derived from the Golgi complex that coalesce during spermiogenesis to form a large, proacrosomal granule. The granule is deposited between the concave surface of the Golgi complex and the nucleus, and the location of the granule determines the future anterior end of the spermatozoon. For more detailed information relevant to the formation and shaping of the acrosome, the works of *Fawcett* (1970), *Plöen* (1971), and *Phillips* (1974) are suggested.

When a mammalian spermatozoon is in the immediate vicinity of the cumulus oophorus, a series of morphological alterations occur on the surface of the acrosomal region. These alterations, known collectively as the acrosome reaction, take place subsequent to the completion of capacitation in species that require this process for fertilization. The occurrence of the acrosome reaction in mammals was first recognized by *Austin* and *Bishop* (1958), and was later more clearly defined by *Austin* (1963). *Austin* observed that spermatozoa with intact acrosomes were unable to move freely within the cumulus mass, in contrast to spermatozoa that had lost the acrosome.

TEM analyses revealed that the acrosome reaction consists of multiple-point fusions (by vesiculation and/or vacuolation) of the outer acrosomal membrane with the overlying plasma membrane (*Pikó* and *Tyler*, 1964; *Barros, Bedford, Franklin*, and *Austin*, 1967; *Bedford*, 1967, 1970; *Jones*, 1973; *Roomans* and *Afzelius*, 1975). The resultant formation of vesicles or membranous discontinuities is presumably required for the activation and/or release of the acrosomal contents. Normally, the process of vesiculation extends only to the limits of the acrosomal cap, with the outer acrosomal and plasma membranes fused at the equatorial region in order to ensure the continuity of the posterior plasma membrane and the maintenance of motility (*Hartree*, 1977). The equatorial segment in some species does undergo vesiculation, but usually at a slightly later time, when the spermatozoa are actually penetrating the zona pellucida (*Bedford*, 1967). As noted by *Bedford* (1970), the acrosome reaction is completed during the passage of spermatozoa through the cumulus and coronal cells, and *must* be accomplished during this passage in order to expose the inner acrosomal membrane. The exposure of the inner acrosomal membrane appears to be necessary for spermatozoa to traverse the zona pellucida (Plate 47B and Chapter 5, section 5.1.3). However, as discussed below, this concept may have to be modified in light of more recent observations.

The origin(s) of the stimulus that elicits the acrosome reaction is controversial. Whereas some investigators have proposed that the oocyte and/or the granulosa cell layers that invest the oocyte may be the source (*Yanagamachi*, 1969a, 1969b; *Pavlok* and *McLaren*, 1972), others have suggested that the factor(s) involved in initiating the acrosome reaction may originate elsewhere (*Fraser, Dandekar*, and *Vaidya*, 1971; *Bavister* and *Morton*, 1974).

The presumed functions of the acrosome reaction are the liberation and activation of the acrosomal contents, permitting spermatozoa to first penetrate the cumulus mass by degrading bonds in the extracellular matrix formed primarily between glycoproteins and hyaluronic acid, and then to traverse the zona pellucida by breaking protein-carbohydrate linkages. Biochemical analyses of isolated mammalian acrosomes demonstrate the presence of both hyaluronidase (*Gould* and *Bernstein*, 1975) and hydrolytic enzymes (*McRorie* and *Williams*, 1974). The hydrolytic enzyme that has been the focus of extensive experimental study is a serine-proteinase that has trypsin-like activity and seems to be required for the digestion of the zona pellucida in some mammals (*Stambaugh* and *Smith*, 1974). This particular enzyme has been termed "acrosin" by *Zaneveld, Robertson, Kessler*, and *Williams* (1971). Other enzymes identified in the acrosomal region include carbonic anhydrase, catalase, lactic dehydrogenase (*Stambaugh* and *Buckley*, 1969, 1970; *Stambaugh*, 1972), aryl phosphatase, aryl sulphatase, phospholipase A, and β-N-acetyloglucosaminidase (*Allison* and *Hartree*, 1970; *McRorie* and *Williams*, 1974, *Hartree*, 1977). Biochemical, immunological, and electron microscopic studies demonstrate that while hyaluronidase is distributed within the acrosomal contents, acrosin is most probably localized to the inner acrosomal membrane (see *Morton*, 1977, for a review).

According to the definition of the acrosome reaction, when a spermatozoon is in the immediate vicinity of, or in contact with, the cumulus mass, the acrosome-plasma membrane complex undergoes vesiculation and is subsequently sloughed off. The disruption of the acrosomal and plasma membranes results in the dispersion of hyaluronidase (and other acrosomal enzymes), which in turn causes the dissolution of the hyaluronidase-labile cementing material of the cellular investments of the oocyte (*Talbot* and *Franklin*, 1974a). Hyaluronidase has also been implicated in facilitating the passage of spermatozoa through the zona pellucida by cleaving zona polysaccharides (*Hartree,* 1971; *Morton,* 1977). When the removal of the acrosome is completed, the inner acrosomal membrane is exposed. Bound to this membrane is presumed to be a proteolytic enzyme(s) capable of the digestion of a narrow path through the zona pellucida (*Garner, Easton, Munson,* and *Doane,* 1975; *Morton,* 1975, 1977; *Garner* and *Easton,* 1977; *Schill, Scheluning, Fritz, Wendt,* and *Heimburger,* 1975).

Several investigators have reported that the penetration of the zona pellucida of rabbit, mouse, rat, and hamster oocytes occurs prior to the dispersal of cumulus cells (*Leonard, Perlman,* and *Kurzrok,* 1947; *Austin,* 1948; *Braden, Austin,* and *David,* 1954), but subsequent to the acrosome reaction (*Austin,* 1969; *Yanagimachi* and *Noda,* 1970a, 1970b, 1972). In addition, *Braden* and *Austin* (1954) and many others have observed that in some species only a very small number of spermatozoa (less than ten in the mouse) could be present when the zona was penetrated prior to the dispersal of cumulus cells. This observation raises the questions of whether hyaluronidase is actually needed to disperse the cumulus mass and, also, whether the removal of these cells is required for the spermatozoa to reach the zona surface. On the other hand, compelling evidence from some species supports the participation of hyaluronidase in the dispersion of the cumulus mass as a prerequisite for fertilization. These species include the guinea pig (where hyaluronidase release and the acrosome reaction are concurrent; *Talbot* and *Franklin,* 1974b) and, possibly, the ram, sheep, cow, and bull (*Morton,* 1977). It seems reasonable to conclude that there may be species variation in the requirement for acrosomal hyaluronidase.

The experimental evidence indicating that acrosin is required for passage of spermatozoa through the zona pellucida is suggestive, but not conclusive. For example, while the penetration of a spermatozoon through the zona is usually accomplished in minutes, purified acrosin, derived from several million spermatozoa, is unable to remove the zona after several hours of exposure (*Morton,* 1977). It is likely that the degradation of the proteo-glycan matrix of the zona involves both trypsin-like enzymes as well as other acrosomal enzymes (including hyaluronidase), perhaps acting synergistically. A wide range of proteolytic enzymes is capable of digesting the zona pellucida *in vitro.* Alternatively, in some species, acrosin alone may be all that is needed to digest the zona. Other suggested functions for acrosin include the induction of the acrosome reaction (*Meizel* and *Lui,* 1976), in-

volvement in the initial attachment of a spermatozoon to the zona pellucida (*Fritz, Schiessler,* and *Schleuning,* 1973), and aiding the release of spermatozoan chromatin into the egg cytoplasm at fertilization by the hydrolysis of spermatozoal histones (*Marushige* and *Marushige,* 1975). The fusion of the male gamete and the vitelline membrane occurs at the equatorial segment (see Chapter 5). Therefore, it has been suggested that if acrosin is involved in this process, it should be present in the equatorial region (*Garner, Reamer, Johnson,* and *Lessley,* 1977). Although these authors failed to detect acrosin in the equatorial segment of the spermatozoa of several mammals by immunoperoxidase staining, the negative results could be explained by inadequate penetration of the antiacrosin antibodies or by a lack of cross-reactivity of the antibodies with the enzymatic components of the acrosome.

4.5. Morphological Changes on the Surface of Spermatozoa Following Mating

The observations presented in this section are derived from the examination of rabbit spermatozoa both in specific regions of the reproductive tract and at specific times following coitus. The rabbit was chosen for this study because its spermatozoa require capacitation and because no morphological correlates of this phenomenon have been documented in TEM studies (*Bedford,* 1970; *Bernstein* and *Teichman,* 1972). The results are derived primarily from a study published by *Motta* and *Van Blerkom* (1975), in which approximately 2,000 spermatozoa were examined. In the ensuing years, 3,000 additional spermatozoa have been analyzed, and these observations have both confirmed and extended the original interpretations. In the following paragraphs, morphological alterations of the acrosomal region that are detected by SEM and that occur during the period of time required for capacitation and the attainment of the fertilizable state of the male gamete are discussed. However, it must be kept in mind that after mating, the female reproductive tract contains virtually millions of spermatozoa, of which a large population may be defective and incapable of fertilizing an oocyte. At present, it is not possible to distinguish by SEM spermatozoa that are competent to fertilize and those that are not. Furthermore, depending on the species, the actual number of spermatozoa that could potentially be involved in fertilization (i.e., at the site of fertilization) may be significantly less than 100 (*Austin,* 1969;

Bedford, 1972)! In the following discussion, we have attempted to convey the existence of subtle morphological changes that occur on the acrosomal region of rabbit spermatozoa while they are in specific regions of the female reproductive tract (especially at the site of fertilization) at specific times following coitus. Consequently, any physiological correlates of morphogenetic events that are introduced must be considered tentative and somewhat speculative. The observations, however, are intended to present information that may be of potential value in the continuing analysis of morphophysiological aspects of spermatozoa as they relate to the attainment of the fertilizable state.

The characteristic appearance of rabbit spermatozoa in the vagina between 1 and 5 minutes postcoitum is shown in Plate 48. The paddle-shaped head of the rabbit spermatozoon is approximately $8\,\mu$m long, $4\,\mu$m wide, and $1\,\mu$m thick (*Flechon* and *Bustos-Obregon,* 1974), whereas the tail is approximately $50\,\mu$m in length (*Hartree,* 1977). Typically, the head of a newly ejaculated spermatozoon is quite smooth and free of surface and/or membranous disruptions. Between 10 and 30 minutes postcoitum, some spermatozoa are covered with a granular/filamentous material which is distributed in an apparently random fashion over both head and tail sections (Plate 49A). In the scanning electron micrograph shown in Plate 49A, however, spermatozoa that are completely devoid of granules may also be observed. The granular/filamentous material is probably derived either from the secretions of the vaginal mucosa and/or from the seminal plasma (Plate 48A).

Spermatozoa observed in the uterus at 3.5 hours postcoitum are generally found in association with ciliated cells (Plates 49B, 50, 51). This association is evident not only in the endocervix where ciliated cells are more numerous (Plate 50A), but also, quite often, in areas of the uterus containing few ciliated cells. The close association of the head region of spermatozoa and ciliated cells is present in both the uterus and the oviduct (Plate 52B) for as many as 24 hours following mating. A complement of surface granules is located on the head region of many spermatozoa, but, in contrast to the situation that prevailed in the vagina, the granules are more regular in appearance and are characteristically limited to the surface of the plasma membrane overlying the acrosome. When the interactions between spermatozoa and cilia throughout the uterus are examined at high magnifications, it becomes plausible to suggest that some of these granules originate from the secretory cells of the uterine mu-

cosa and are subsequently accumulated and distributed by the ciliated cells. Upon contact with the ciliated cells, the granular material seems to be deposited on the spermatozoal surface and then localized over the acrosomal cap (Plates 49B, 50) (*Kanagawa* and *Hafez,* 1973). Furthermore, many spermatozoa observed in regions of the uterus in which few ciliated cells are present nevertheless contain a population of surface (acrosomal) granules; perhaps this may indicate an earlier encounter with ciliated cells. Although the precise origin and biochemical composition of the granules are unknown, available histochemical evidence suggests a mucopolysaccharide nature (see Chapter 3 and *Lawn,* 1974, for references).

The uterine epithelia of rats, mice (*Bergström,* 1970, 1972; *Smith* and *Wilson,* 1971; *Moulton,* 1974; *Meglioli* and *Desaulles,* 1975; *Nilsson,* 1977), and rabbits (*Kirchner,* 1972; *Denker* and *Hafez,* 1975) are capable of secreting hydrolytic enzymes, possibly enclosed in small vesicles. In addition, *McLaren* and *Nilsson* (1971) and *Nilsson* (1977) reported that the qualitative and quantitative character of the secretions were functions of the size and number of luminal objects, such as preimplantation embryos. What is of direct interest to the topic under consideration is that luminal objects could elicit a local secretory response from the uterine epithelium. Unfortunately, it is not currently known whether spermatozoa are capable of eliciting such a response, or whether the surface granules on spermatozoa represent residual plasma proteins, mucopolysaccharides, and/or hydrolytic enzymes enclosed in vesicles. Further studies are required to establish not only the chemical nature of the granules, but also whether they may contain lytic substances that could possibly function in the labilization of the acrosomal surface (*Motta* and *Van Blerkom,* 1975). Suggestive of a lytic function is the finding that while the plasma membrane overlying the acrosome is free of disruption in the majority of spermatozoa examined at 3.5 hours postcoitum, some spermatozoa do exhibit slight membranous perforations, localized in regions adjacent and subjacent to the granules. Collectively, the above observations suggest two potentially significant avenues of investigation relating to the post-ejaculatory distribution and/or modification of antigenic and lectin-binding sites on the acrosome (see section 4.1; *Johnson,* 1975; *Johnson* and *Howe,* 1975): (1) does the apparent localization of granules to the acrosomal cap involve specific interactions between acrosomal macromolecules and the granules, and (2) are specific binding sites made available by exposure of spermatozoa to the physical and chemical milieu of the female reproductive tract?

Although the plasma membrane that overlies the acrosome is intact in some spermatozoa at 9 hours postcoitum (Plate 51A), a sizable population of spermatozoa display varying degrees of surface and membranous alterations in this region (Plates 51B, 52A, 53). A dynamic reconstruction of the origin of these disruptions, derived from an extensive examination of a large population of spermatozoa, provides evidence in support of the following sequence of events. At the equatorial segment, a series of small fenestrations in the plasma membrane arises by what appears to be a process of vesiculation. With time, these small perforations coalesce, and larger areas of the plasma membrane are gradually disrupted. At high SEM magnification, it is apparent that it is the plasma membrane (and possibly also a glycoprotein coat) that is being eroded. It is also evident at high magnification that for some spermatozoa, the erosion process includes portions of the outer acrosomal membrane as well. Typically, the disruption of the membranous coverings is limited to the acrosomal region, since no sign of alteration is observed in either the postacrosomal zone (Plates 51, 52, 53) or the plasma membrane that covers the tail. In cases of advanced disruption, the region of the acrosome formerly occupied by the intact plasma membrane contains large numbers of vesicle-like structures (Plate 53C). These structures most likely correspond to the vesicles observed by TEM and thought to be associated with the acrosome reaction (*Pikó* and *Tyler*, 1964; *Barros, Bedford, Franklin,* and *Austin,* 1967). It is also likely that these vesicles are transitory in nature and after their formation do not adhere to spermatozoa for prolonged periods of time.

A small number of spermatozoa at 9 hours postcoitum appears to have unaltered plasma membranes, although a complement of surface granules is present (Plate 51A). On the other hand, the same relative proportion of spermatozoa observed both throughout the oviduct (especially in the reduced population of spermatozoa at the site of fertilization) and on the surface of the ovary at 9 hours postcoitum display the same range of acrosomal alterations as described above (*Motta* and *Van Blerkom,* 1975). Because approximately 9 hours are required to capacitate rabbit spermatozoa, these observations necessarily raise questions concerning the potential existence of morphological correlates of this process. The implications of these results are discussed below.

Between 15 and 24 hours postcoitum, the surface of the majority of spermatozoa located in the upper regions of the uterus, throughout the oviduct, and on the surface of the ovary and/or the cumulus oophorus of an ovulated oocyte is extensively disrupted. In some cases, portions of the membranous coverings of the acrosome appear to be exfoliating (Plate 54). This exfoliation is generally limited to spermatozoa that have resided in the female reproductive tract for more than 15 hours and appears to be a distinctly different process from the membranous disruptions observed at earlier times following mating. In the majority of the spermatozoa that display the typical pattern of surface disruption, numerous vesicle-like structures are visible over the acrosome (Plates 53C, 55A). However, even after as many as 24 hours, some spermatozoa are encountered in which no obvious disruption/vesiculation of the membranous coverings of the acrosome has taken place (Plate 54C). In addition, at 24 hours postcoitum, many spermatozoa are found associated with, in close proximity to, or engulfed by leukocytes (Plates 55B, 55C). By 36 hours, only a reduced population of spermatozoa is available for analysis, and most of these cells exhibit extensive degeneration of both head and tail sections (Plate 55D).

At present, it is rather difficult to reconcile images obtained by SEM with the traditional view of capacitation and the acrosome reaction as defined by TEM (see *Bedford,* 1970). The collective results of the SEM examination of approximately 5,000 rabbit spermatozoa indicate that spermatozoa exist in the reproductive tract of the female as a morphologically heterogeneous population and that shifts within this population from one predominant surface morphology to another take place with time. The pattern of alteration of the acrosomal region represents a time-dependent labilization of the plasma membrane and any surface coat (and possibly the outer acrosomal membrane as well), which characteristically is initiated at the equatorial segment as a series of small fenestrations. With increasing intervals of time, the fenestrations coalesce, with the result that progressively larger areas of the membranes and cell coat that overlie the acrosome are disrupted. In numerous spermatozoa observed at 9 hours postcoitum, the surface and membranous disruptions are well advanced, and vesicle-like structures are encountered over the acrosomal region but, significantly, not over the postacrosomal zone. Many of the alterations described above rather closely resemble morphological changes observed by TEM and termed the "acrosome reaction" (or "true acrosome reaction"), but only if occurring in the immediate vicinity of an oocyte (*Bed-*

ford, 1970). It is worthwhile to recall that by TEM, the acrosome reaction consists of the fusion (at various points) of the outer acrosomal and plasma membranes through a process of vesiculation and/or vacuolization.

One explanation of the morphological sequence of events detectable by SEM is that the disrupted spermatozoa are actually undergoing "premorbid" changes, which, under physiological conditions, render them incapable of fertilizing an oocyte (i.e., passing through the cellular and noncellular layers of an oocyte). If this interpretation is correct, then the surface and/or membranous alterations may merely represent normal degenerative processes. In this regard, *Bedford* (1970) has described the false acrosome reaction as one in which alterations affect the plasma membrane of a spermatozoon that is removed from the immediate vicinity of an oocyte. However, detailed TEM investigations have failed to detect morphological differences between the false and true acrosome reaction (which occurs near an oocyte) (*Zamboni,* 1971, 1972), although a precocious acrosome reaction could have physiological consequences relevant to the ability of a spermatozoon to penetrate and fertilize an oocyte (*Bedford,* 1970, 1972).

Another interpretation of the SEM evidence is that the surface and/or membranous alterations are intimately involved in the attainment of the fertilizable state of the male rabbit gamete. That a progressively increasing population of spermatozoa display similar patterns of time-dependent disruptions over the acrosome, especially in spermatozoa located at the site of fertilization, could tentatively support a morpho-physiological process related to capacitation and/or the acrosome reaction. Furthermore, these subtle changes might not be detected by TEM. Although it is premature to conclude that a functional correspondence exists between the disruption of plasma and acrosomal membranes and capacitation and/or the acrosome reaction, it is nevertheless an intriguing possibility that such a relationship exists. In this respect, the concept of capacitation and the acrosome reaction, rather than being viewed as distinctly separate but related phenomena, may be considered in a unified context (*Williams,* 1972). Thus, capacitation would include the gradual labilization of the plasma membrane (and removal of any surface coat) leading to the acrosome reaction, which is initiated as small perforations at the equatorial segment and consists of the progressive exposure of limited regions of the acrosomal contents to the milieu of the reproductive tract. This process would ultimately lead to the dis-

persion of acrosomal enzymes and, in some species, would coincide in time with the presence of an ovulated oocyte.

The observation that the disruption of the plasma membrane is *specifically* limited to the acrosome indicates not only that the fragility of this membrane may be related to the lytic activity of acrosomal enzymes, but also supports the hypothesis that this particular region has a specific macromolecular and structural organization that may be sensitive to conditions or factors present within the female reproductive tract. The only experimental evidence in favor of this interpretation at present is the finding that *in vitro* capacitated hamster spermatozoa release hyaluronidase (and possibly other acrosomal enzymes) several hours *before* undergoing an obvious acrosome reaction and that this release is unrelated to cell death or degeneration (*Rogers* and *Morton,* 1973; *Talbot* and *Franklin,* 1974c). Thus, it is possible that exposure of spermatozoa to the physical and chemical environment of the female reproductive tract could initiate a gradual release or activation of acrosomal enzymes prior to the appearance of a definitive acrosome reaction, and such activity could be related to subtle morphological changes visible by SEM.

Clearly, much more experimental evidence must be accumulated in order to reconcile these contrasting interpretations. Spermatozoa of other species, especially those requiring capacitation as well as those that definitely do not, must be subjected to comparative TEM and SEM analysis to determine (1) whether similar modifications to the cell coat and/or plasma membrane occur, (2) whether such changes are dependent on time and/or location in the female reproductive tract, and, perhaps of greatest importance, (3) whether alterations of this nature are indeed related to fertilization. The basic conclusion that may be derived from this section is that, at least in the rabbit, spermatozoa exist in the female reproductive tract as a morphologically heterogeneous population, and that, with time, significant alterations in surface architecture occur on the acrosome. These processes may be important in terms of the attainment of the fertilizable state of the male gamete. Alternatively, they may merely represent random degenerative processes. However, the concept of a population of spermatozoa should be kept in mind when biochemical, immunological, and morphological approaches are used to investigate and elucidate the processes by which a spermatozoon becomes competent to fertilize an oocyte.

References

Allison, A.C., and *Hartree, E.F.* (1970). Lysosomal enzymes in the acrosomes and their possible role in fertilization. J. Reprod. Fert. *21,* 501–515.

Austin, C. R. (1948). Function of hyaluronidase in fertilization. Nature (London) *162,* 63–64.

Austin, C. R. (1951). Observations of the penetration of sperm into the mammalian egg. Aust. J. Sci. Res. B. *4,* 581–589.

Austin, C. R. (1952). The capacitation of the mammalian spermatozoa. Nature (London) *170,* 326.

Austin, C. R. (1963). Acrosome loss from the rabbit spermatozoon in relation to entry into the egg. J. Reprod. Fert. *6,* 313–314.

Austin, C. R. (1969). Sperm capacitation—biological significance in various species. Advances in Biosciences, *6.*

Austin, C. R. (1975). Membrane fusion events in fertilization. J. Reprod. Fert. 44, 155–166.

Austin, C. R., and *Bishop, M. W. H.* (1958). Role of the rodent acrosome and perforation in fertilization. Proc. Royal Soc. B. *149,* 241–248.

Baccetti, B. (1970). Comparative Spermatology. Academic Press, New York.

Baccetti, B., and *Afzelius, B. A.* (1976). The biology of the sperm cell. In: Monographs in Developmental Biology, Vol. 10 (Welsky, A., ed.). S. Karger, Basel.

Barros, C., Bedford, J. M., Franklin, M. C., and *Austin, C. R.* (1967). Membrane vesiculation as a feature of the mammalian acrosome reaction. J. Cell Biol. *34,* C1–C5.

Barros, C., and *Garavango, A.* (1970). Capacitation of rabbit spermatozoa with blood sera. J. Reprod. Fert. *22:*381–384.

Barros, C., Berrios, M., and *Herrera, E.* (1973). Capacitation *in vitro* of guinea-pig spermatozoa in a saline solution. J. Reprod. Fert. *34 :* 547–549.

Bavister, B. D. (1973). Capacitation of golden hamster spermatozoa during incubation in culture medium. J. Reprod. Fert. *35:*161–163.

Bavister, B. D., and *Morton, D. B.* (1974). Separation of human serum components capable of inducing the acrosome reaction in hamster spermatozoa. J. Reprod. Fert. *40:*495–498.

Bedford, J. M. (1967). Experimental requirement for capacitation and observations on ultrastructural changes in rabbit spermatozoa during fertilization. J. Reprod. Fert., suppl. *2,* 35–48.

Bedford, J. M. (1970). Sperm capacitation and fertilization in mammals. Biol. Reprod., suppl. *2,* 128–158.

Bedford, J. M. (1972). Sperm transport, capacitation and fertilization. In: Reproductive Biology (H. Balin and S. Glasser, eds.). Excerpta Medica, Amsterdam.

Bergström, S. (1970). Estimation of proteolytic activity at mouse implantation sites by the gelatin digestion method. J. Reprod. Fert. *23:*481–485.

Bergström, S. (1972). Histochemical localization of acid uterine aminoacylnaphthylamidases in early pregnancy and in different hormonal states of the mouse. J. Reprod. Fert. *30:*177–183.

Bernstein, M. H., Teichman, R. J. (1972). Morphological aspects of capacitation. In: Biology of Mammalian Fertilization and Implantation, pp. 126–138 (K. S. Moghissi and E.S.E. Hafez, eds.). Charles C. Thomas, Springfield, Ill.

Brackett, B. G., and *Oliphant, G.* (1975). Capacitation of rabbit spermatozoa *in vitro.* Biol. Reprod. *12:*260–274.

Braden, A. W. H., and *Austin, C. R.* (1954). The number of sperma about the eggs in mammals and its significance for normal fertilization. Aust. J. Biol. Sci. *1:*543–551.

Braden, A. W. H., Austin, C. R., and *David, H. A.* (1954). The reaction of the zona pellucida to sperm penetration. Aust. J. Biol. Sci. 7:391–409.

Bryan, J. H. D. (1974). Capacitation in the mouse: The response of murine acrosomes to the environment of the female reproductive tract. Biol. Reprod. *10:*414–421.

Chang, M. C. (1951). Fertilizing capacity of spermatozoa deposited into the Fallopian tubes. Nature (London) 168:697.

Denker, H.-W., and *Hafez, E. S. E.* (1975). Proteases and implantation in the rabbit: Role of trophoblast vs. uterine secretion. Cytobiologie *11:*101–109.

Fawcett, D. W. (1970). A comparative view of sperm ultrastructure. Biol. Reprod. *2:*90–127.

Flechon, J. E., and *Bustos-Obregon, E.* (1974). Scanning electron microscope study of rabbit spermatozoa. Andrologia 6:169–180.

Fraser, L. R., Dandekar, P. V., and *Vaidya, R. A.* (1971). *In vitro* fertilization of tubal rabbit ova partially or totally denuded of follicular cells. Biol. Reprod. *4:*229–233.

Friend, D. S. (1977). The organization of the spermatozoal membrane. In: Immunobiology of the Gametes (M. Edidin and M. H. Johnson, eds.). Cambridge University Press, Cambridge, England.

Friend, D. S., and *Fawcett, D. W.* (1974). Membrane differentiation in freeze-fractured mammalian sperm. J. Cell Biol. *63:*641–664.

Fritz, H., Schiessler, H., and *Schleuning, W. D.* (1973). Proteinases and proteinase inhibitors in the fertilization process: New concepts of control. Advances in the Biosciences *10:*271–284.

Garner, D. L., and *Easton, M. P.* (1977). Immunofluorescent localization of acrosin in mammalian spermatozoa. J. Exp. Zool. *200:*157–162.

Garner, D. L., Easton, M. P., Munson, M. E., and *Doane, M. A.* (1975). Immunofluorescent localization of bovine acrosin. J. Exp. Zool. *191:*127–131.

Garner, D. L., Reamer, S. A., Johnson, L. A., and *Lessly, B. A.* (1977). Failure of immunoperoxidase staining to detect acrosin in the equatorial segment of spermatozoa. J. Exp. Zool. *201:*309–315.

Gordon, M., Dandekar, P. V., and *Bartoszewicz, W.* (1975). The surface coat of epididymal, ejaculated and capacitated sperm. J. Ultr. Res. *50:*199–207.

Gould, S. F., and *Bernstein, M. H.* (1975). The localization of bovine sperm hyaluronidase. Differentiat. *3:*123–132.

Gwatkin, R. B. L., and *Hutchison, C. F.* (1971). Capacitation of hamster spermatozoa by β-glucuronidase. Nature (London) 229:343–344.

Hartree, E. F. (1971). Lysosomes and fertilization. In: Of Microbes and Life, pp. 271–303 (J. Monod and E. Borek, eds.). Columbia University Press, New York and London.

Hartree, E. F. (1975). The acrosome-lysosome relationship. J. Reprod. Fert. *44:*125–126.

Hartree, E. F. (1977). Spermatozoa, eggs and proteinases. Biochem. Soc. Trans. *5:*375–394.

Johnson, M. H. (1975). The macromolecular organization of membranes and its bearing on events leading up to fertilization. J. Reprod. Fert. *44*:167–184.

Johnson, M. H., and Howe, C. W. S. (1975). The immunobiology of spermatozoa. In: The Biology of the Male Gamete (J. G. Duckett and P. A. Racey, eds.). Biol. J. Linnean Soc. 7, suppl. *1*:205–214.

Jones, R. C. (1973). Changes occurring in the head of boar spermatozoa: Vesiculation or vacuolation of the acrosome? J. Reprod. Fert. *33*:113–118.

Kanagawa, H., and Hafez, E. S. E. (1973). Kinocilia and sperm dynamics in the cervix uteri of the rabbit. J. Reprod. Med. *10*:90–94.

Kirchner, C. (1972). Uterine protease activities and lysis of the blastocyst covering in the rabbit. J. Embryol. exp. Morph. *28*:177–183.

Koehler, J. K. (1975). Studies on the distribution of antigenic sites on the surface of rabbit spermatozoa. J. Cell Biol. *67*:647–659.

Koehler, J. K., and Perkins, W. D. (1974). Fine structure observations on the distribution of antigenic sites on guinea pig spermatozoa. J. Cell Biol. *60*:789–795.

Koo, G. C., Boyse, E. A., and Wachtel, S. S. (1977). Immunogenetic techniques and approaches in the study of sperm and testicular cell surface antigens. In: Immunobiology of Gametes (M. Edidin and M. H. Johnson, eds.). Cambridge University Press, Cambridge, England.

Lawn, A. M. (1974). The ultrastructure of the endometrium during the sexual cycle. In: Advances in Reproductive Physiology, Vol. 6 (M. N. H. Bishop, ed.). Elek. Science, London.

Leonard, S. L., Perlman, P. L., and Kurzrok, R. (1947). Relation between time of fertilization and follicle cell dispersal in rat ova. Proc. Soc. Experimental Biol. and Medicine *66*:517–518.

Marushige, Y., and Marushige, K. (1975). Enzymatic unpacking of bull sperm chromatin. Biochim. et Biophys. Acta *403*:180–191.

McLaren, A., and Nilsson, O. (1971): Electron microscopy of luminal epithelium separated by beads in the pseudopregnant mouse uterus. J. Reprod. Fertil. 26, 379–381.

McRorie, R. A., and Williams, W. L. (1974). Biochemistry of mammalian fertilization. Ann Rev. Biochem. *43*:777–803.

Meglioli, G., Desaulles, P. A. (1975). Uterine secretion during the sexual cycle in the rat and its capacity to disperse corona cells *in vitro*. Experientia *31*:986–989.

Meizel, S., and Lui, C. W. (1976). Evidence for the role of a trypsin-like enzyme in the hamster sperm acrosome reaction. J. Exp. Zool. *195*:137–144.

Millette, C. F. (1977). Distribution and motility of lectin binding sites on mammalian spermatozoa. In: Immunobiology of Gametes (M. Edidin and M. H. Johnson, eds.). Cambridge University Press, Cambridge, England.

Millette, C. F., and Bellve, A. R. (1977). Temporal expression of membrane antigens during mouse spermatogenesis. J. Cell Biol. *74*:86–97.

Morton, D. B. (1975). Acrosomal enzymes: Immunochemical localization of acrosin and hyaluronidase in ram spermatozoa. J. Reprod. Fert. *45*:375–378.

Morton, D. B. (1977). Immunoenzymatic studies on acrosin and hyaluronidase in ram spermatozoa. In: Immunobiology of Gametes (M. Edidin and M. H. Johnson, eds.).

Cambridge University Press, Cambridge, England.

Motta, P., and Van Blerkom, J. (1975). A scanning electron microscopic study of rabbit spermatozoa in the female reproductive tract following coitus. Cell Tiss. Res. *163*:29–44.

Moulton, B. C. (1974). Ovum implantation and uterine lysosomal enzyme activity. Biol. Reprod. *10*:543–548.

Nilsson, O. (1977). Local secretory response in the mouse uterine epithelium to the presence of a blastocyst or a blastocyst-like bead. Anat. Embryol. *150*:313–318.

Oliphant, G., and Brackett, B. G. (1973). Capacitation of mouse spermatozoa in media with elevated ionic strength and reversible decapacitation with epididymal extracts. Fert. Steril. *24*:948–955.

Pavlok, A., and McLaren, A. (1972). The role of cumulus cells and the zona pellucida in fertilization of mouse eggs *in vitro*. J. Reprod. Fert. *29*:91–97.

Phillips, D. M. (1974). Spermiogenesis. Academic Press, New York.

Pikó, L., and Tyler, A. (1964). Fine structural studies of sperm penetration in the rat. Proc. 5th Internat Cong. Anim. Reprod. & A. I. Trento *2*:372–377.

Plöen, L. (1971). A scheme of rabbit spermateleosis based upon electron microscopic observations. Z. Zellforsch. mikrosk. Anat. *115*:553–564.

Rogers, B. J., and Morton, B. E. (1973). The release of hyaluronidase from capacitating hamster spermatozoa. J. Reprod. Fert. *35*:477–487.

Roomans, G. M., and Afzelius, B. A. (1975). Acrosome vesiculation in human sperm. J. Submicr. Cytol. *7*:61–70.

Schill, W.-B., Scheluning, W.-D., Fritz, H., Wendt, V., and Heimburger, N. (1975). Immunofluorescent localization of acrosin in spermatozoa by boar acrosin antibodies. Die Naturwissenschaften *11*:540.

Smith, M. S. R., and Wilson, I. B. (1971). Histochemical observations on early implantation in the mouse. J. Embryol. exp. Morph. *25*:165–174.

Soupart, P. (1972). Sperm capacitation: Methodology, hormonal control and the search for a mechanism. In: Biology of Mammalian Fertilization and Implantation (K. S. Moghissi and E. S. E. Hafez, eds.). Charles C. Thomas, Springfield, Ill.

Stackpole, C. W., and Devorkin, D. (1974). Membrane organization in mouse spermatozoa revealed by freeze-etching. J. Ultr. Res. *49*:167–187.

Stambaugh, R. L. (1972). Acrosomal enzymes and fertilization. In: Biology of Mammalian Fertilization and Implantation (K. S. Moghissi and E. S. E. Hafez, eds.). Charles C. Thomas, Springfield, Ill.

Stambaugh, R., and Buckley, J. (1969). Identification and sub-cellular localization of the enzymes affecting penetration of the zona pellucida by rabbit spermatozoa. J. Reprod. Fert. *19*:423–432.

Stambaugh, R., and Buckley, J. (1970). Comparative studies of the acrosomal enzymes of rabbit, rhesus monkey and human spermatozoa. Biol. Reprod. *3*:275–282.

Stambaugh, R., and Smith, M. (1974). Amino acid content of rabbit acrosomal proteinase and its similarity to human trypsin. Science *186*:745–746.

Talbot, P., and Franklin, L. E. (1974a). The release of hyaluronidase from guinea-pig spermatozoa during the course of the normal acrosome reaction *in vitro*. J. Reprod. Fert. *39*:429–432.

Talbot, P., and *Franklin, L. E.* (1974b). Hamster sperm hyaluronidase. A bioassay procedure based on cumulus dispersion rate. J. Exp. Zool. *189*:311–320.

Talbot, P., and *Franklin, L. E.* (1974c). Hamster sperm hyaluronidase. Its release from sperm *in vitro* in relation to the degenerate and normal acrosome reaction. J. Exper. Zool. *189*:321–332.

Williams, W. L. (1972). Biochemistry of capacitation of spermatozoa. In: Biology of Mammalian Fertilization and Implantation (K. S. Moghissi and E. S. E. Hafez, eds.). Charles C. Thomas, Springfield, Ill.

Yanagimachi, R. (1969a). *In vitro* capacitation of hamster spermatozoa by follicular fluid. J. Reprod. Fert. *18*:275–286.

Yanagimachi, R. (1969b). *In vitro* acrosome reaction and capacitation of golden hamster spermatozoa with bovine follicular fluid and its fractions. J. Exper. Zool. *170*:269–280.

Yanagimachi, R. (1970a). *In vitro* capacitation of golden hamster spermatozoa by homologous and heterologous blood sera. Biol. Reprod. *3*:147–153.

Yanagimachi, R., and *Noda, Y. D.* (1970a). Ultrastructural changes in the hamster sperm head during fertilization. J. Ultr. Res. *31*:465–485.

Yanagimachi, R., and *Noda, Y. D.* (1970b). Physiological changes in the postnuclear cap region of mammalian spermatozoa; a necessary preliminary to the membrane fusion between sperm and egg cells. J. Ultr. Res. *31*:486–493.

Yanagimachi, R., and *Noda, Y. D.* (1972). Scanning electron microscopy of golden hamster spermatozoa before and during fertilization. Experientia *28*:69–72.

Zamboni, L. (1971). The fine Morphology of Mammalian Fertilization. Harper & Row, New York.

Zamboni, L. (1972). Fertilization in the mouse. In: Biology of Mammalian Fertilization and Implantation (K. S. Moghissi and E. S. E. Hafez, eds.). Charles C. Thomas, Springfield, Ill.

Zaneveld, L. J. D., Robertson, R. T., Kessler, M., and *Williams, W. L.* (1971). Inhibition of fertilization *in vivo* by pancreatic and seminal plasma trypsin inhibitors. J. Reprod. Fert. *25*:387–392.

Plate 47. General Organization of Mammalian Spermatozoa.

A   Diagrammatic representation of a typical mammalian spermatozoon. The various regions and components of the spermatozoon illustrated here have been defined by transmission electron microscopy.

B   Transmission electron micrograph of a mouse spermatozoon. Ac = acrosome, N = nucleus, PM = plasma membrane, AM = acrosomal membranes, Sp = spermatocytes. Sertoli cell projections and/or debris = arrows*. (x 18,000).

C   High-magnification transmission electron micrograph of the middle piece of a mouse spermatozoon. M = mitochondria, Mt = microtubules, Df = dense fibers. (x 40,000).

146

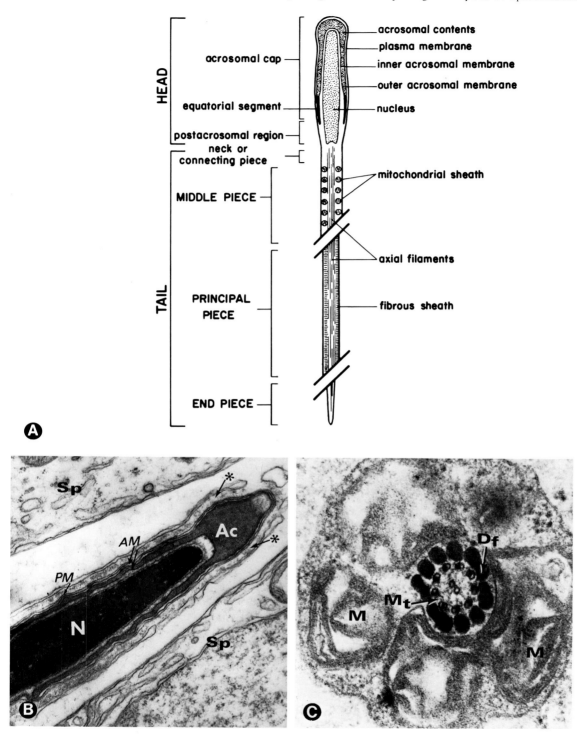

Plate 47. General Organization of Mammalian Spermatozoa.

Plate 48. The Appearance of Rabbit Spermatozoa in the Vagina Between 1 and 5 Minutes Postcoitum.

A   Typically, the acrosomal region of freshly ejaculated spermatozoa is free of surface and/or membranous disruptions. In this scanning electron micrograph, a small population of surface granules (arrow) as well as strands of an amorphous material (arrow*) which probably originated from surrounding cells of the vaginal mucosa are evident on the acrosomal surface (Ac). (x 13,700).

B   The characteristic "paddle-shaped" head of a rabbit spermatozoon is demonstrated in this scanning electron micrograph. The surface of the acrosomal cap (Ac) contains granules and accumulations of an amorphous material (arrows). Notice in particular the mucus-like material present on the microvilli (Mv) of the vaginal mucosa. T = tail. (x 14,000).

Plate 48. The Appearance of Rabbit Spermatozoa in the Vagina Between 1 and 5 Minutes Postcoitum.

Plate 49. Appearance of Rabbit Spermatozoa in the Vagina and Uterus.

A   Between 10 and 30 minutes postcoitum, some spermatozoa were covered to varying degrees with a granular material that was distributed over both head and tail regions in an apparently random fashion (arrows). However, spermatozoa were observed in which no surface granules were present (*). (x 6,400; vagina).

B   Scanning electron micrograph of a spermatozoon in the uterus at 3.5 hours postcoitum. At this time, most spermatozoa contained a complement of surface granules (arrows) localized in the acrosomal cap (Ac) and which appeared to have been transferred from surrounding cilia (C). PA = postacrosomal region, T = tail. (x 17,860).

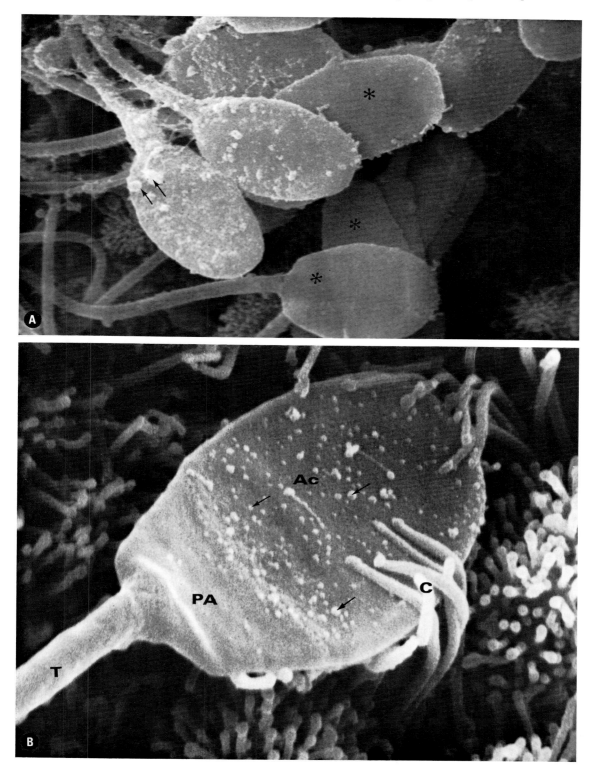

Plate 49. Appearance of Rabbit Spermatozoa in the Vagina and Uterus.

Plate 50. The Appearance of Rabbit Spermatozoa in the Uterus at 3.5 Hours Postcoitum.

A   At 3.5 hours postcoitum, many spermatozoa were found in association with ciliated cells of the uterus. Sc = secretory cell, Cc = ciliated cell. (x 3,800).

B   Although the surface of most spermatozoa appeared undisrupted at 3.5 hours postcoitum, some spermatozoa did contain regions overlying the acrosomal cap in which a relatively mild disruption of either a glycocalyx coat and/or the plasma membrane overlying the acrosome had taken place (arrows, *). C = cilia, Sc = secretory cell. (x 8,000).

C   The intimate association between the acrosomal region of a spermatozoon and cilia is shown in this scanning electron micrograph. At high magnification, it appears as if the granules located on the acrosomal region were transferred from the endometrium by the beating and brushing action of cilia (arrows). (x 20,000).

Plate 50. The Appearance of Rabbit Spermatozoa in the Uterus at 3.5 Hours Postcoitum.

Plate 51. The Appearance of Rabbit Spermatozoa at 9 Hours Postcoitum.

A   At 9 hours postcoitum, a small but significant population of spermatozoa displayed intact surfaces, although a population of granules (arrows) was present on the acrosomal region (Ac). At high magnification, areas of limited surface erosion were observed in the portion of the plasma membrane subjacent to the granules (arrows*). Spermatozoa of the same morphological appearance were encountered at this time throughout the oviduct and on the surface of the ovary. PA = postacrosomal region. (x 19,000).

B   In a much greater proportion of spermatozoa than observed previously, the portion of the head that was anterior to the equatorial segment (ES) displayed varying degrees of disruption (arrows). The extent of disruption shown in this scanning electron micrograph was relatively mild in comparison to the superficial appearance of many other spermatozoa (see Plate 52 A) observed in the uterus, throughout the oviduct, and on the surface of the ovary at 9 hours postcoitum. C = cilia. (x 18,800).

154

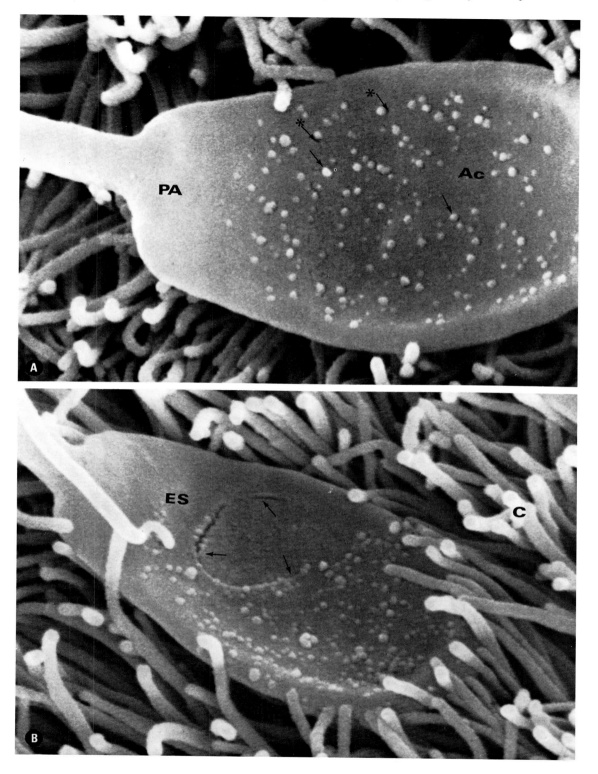

Plate 51. The Appearance of Rabbit Spermatozoa at 9 Hours Postcoitum.

Plate 52. The Appearance of Rabbit Spermatozoa at 9 Hours Postcoitum.

A   Although the spermatozoa shown in this scanning electron micrograph were located in the uterus, the surface disruption of the acrosomal region was characteristic of spermatozoa throughout the oviduct and on the surface of the ovary. The degree of disruption demonstrated by these spermatozoa is fairly extensive, although it is not clear whether the disruption involves a surface coat and/or the membranes that overlie the acrosome. Notice in particular that the erosion of the surface is limited to the portion of the head anterior to the equatorial segment (ES, arrows). C = cilia, PA = postacrosomal region. (x 17,800).

B   Scanning electron micrograph illustrating the association of spermatozoa and ciliated cells (C). Sc = secretory cells. (x 3,800).

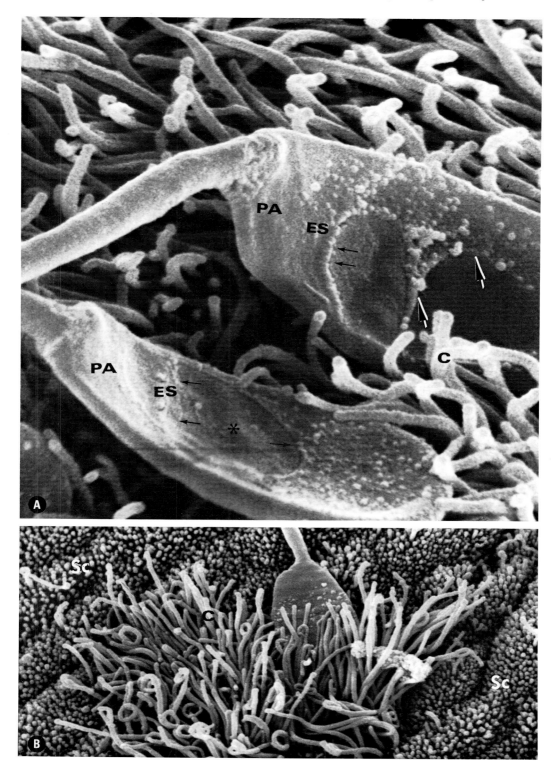

Plate 52. The Appearance of Rabbit Spermatozoa at 9 Hours Postcoitum.

Plate 53. Appearances of Rabbit Spermatozoa at 9 Hours Postcoitum.

A   The appearance of a spermatozoon on the surface of a preovulatory follicle at 9 hours postcoitum is demonstrated in this scanning electron micrograph. The disruption of the surface of the acrosomal region is characteristically limited to the portion of the head anterior to the equatorial segment (*). At this time, most of the spermatozoa observed throughout the oviduct or located on preovulatory follicles displayed this surface morphology. (x 10,000).

B   At 9 hours postcoitum, numerous granules and/or vesicle-like structures were present on the acrosomal region of many spermatozoa (arrows) located in the upper regions of the oviduct. (x 15,000).

C   In the ampullary region of the oviduct at 9 hours postcoitum, numerous spermatozoa were encountered in which large areas of the plasma membrane that overlie the acrosomal cap were missing. Instead, these areas were populated with vesicle-like structures (arrows) reminiscent of the vesicles observed by TEM and formed between the plasma membrane and outer acrosomal membrane during the acrosome reaction. Again, notice that the disruption occurs anterior to the equatorial segment (ES) and that the postacrosomal region (PA) is apparently undisturbed. (x 8,000).

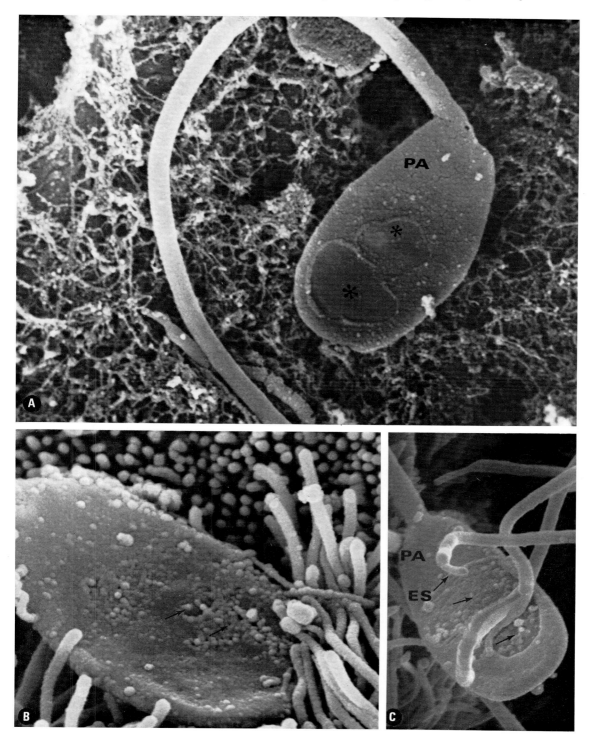

Plate 53. Appearance of Rabbit Spermatozoa at 9 Hours Postcoitum.

Plate 54. Appearance of Rabbit Spermatozoa at 15 Hours Postcoitum.

A   At 15 hours postcoitum, the acrosomal region of most spermatozoa, present either in the uterus, oviduct, or on the surface of the ovary and/or the cumulus mass of an ovulated oocyte, displayed extensive surface and membranous disruptions. In this scanning electron micrograph, the membranes covering the acrosomal region have been extensively eroded owing to the prior coalescence of small fenestrations that originated anterior to the equatorial segment (ES). (x 10,900).

B   Occasionally, spermatozoa were observed in which large areas of the plasma membrane overlying the acrosome (Ac) appeared to be exfoliating (arrows). This morphology was relatively rare and may be related to a process of degeneration rather than to an event involved in the attainment of the fertilizable state. (x 11,200).

C   Even after as many as 15 hours postcoitum, some spermatozoa were present in which no apparent surface or membranous disruption had occurred (spermatozoon Ac). The more common morphology that was observed is illustrated by the spermatozoon at the lower right-hand portion of the scanning electron micrograph, in which a large area of the plasma membrane overlying the acrosome (Ac*) has been eroded. Notice the presence of relatively large secretory droplets (arrows) on the surface of the uterine mucosa and the association of cilia (C) with the head region of the spermatozoa. Sc = secretory cell, S = microvillous cell with an isolated cilium. (x 5,700).

160

Plate 54. Appearance of Rabbit Spermatozoa at 15 Hours Postcoitum.

Plate 55. Appearance of Rabbit Spermatozoa at 15, 24, and 36 Hours Postcoitum.

A   At 24 hours postcoitum, nearly all spermatozoa examined by SEM (throughout the entire reproductive tract) displayed extensive disruption of the acrosomal region (*). However, an occasional spermatozoon was encountered in which either no apparent disruption had taken place, or the disruption was relatively slight (spermatozoon with two arrows). Secretory droplets on the surface of the uterine mucosa were also observed on the head and tail sections of some spermatozoa at this time (arrows). C = cilia. (x 2,300).

B   At 15 hours postcoitum, spermatozoa were observed in association with leukocytes (Lc). Arrows indicate the equatorial segment. (x 4,800).

C   Scanning electron micrograph of spermatozoa in the uterus at 15 hours postcoitum demonstrating the close association of leukocytes (Lc) and spermatozoa. (x 1,950).

D   By 36 hours postcoitum, almost every spermatozoa examined was in an advanced stage of degeneration. The degeneration process encompassed both the head (H) and tail (T) regions. (x 7,300).

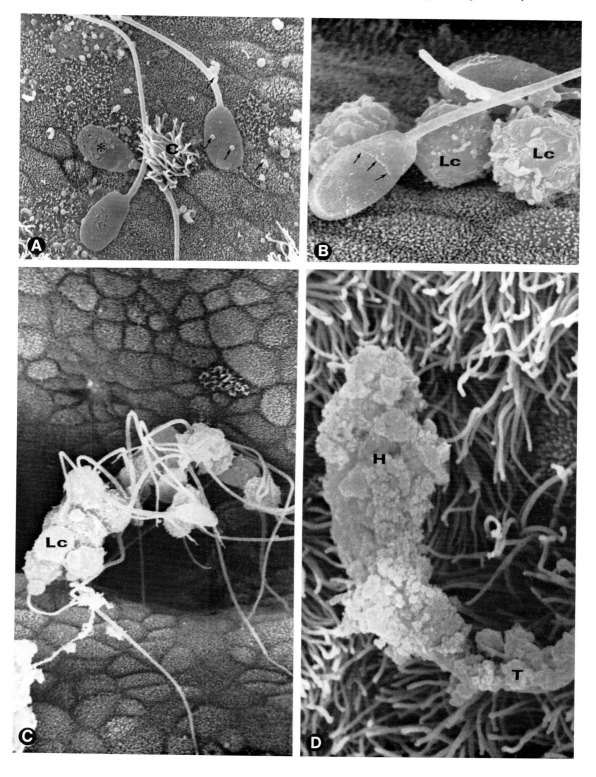

Plate 55. Appearance of Rabbit Spermatozoa at 15, 24, and 36 Hours Postcoitum.

5 Fertilization and Preimplantation Embryogenesis

5.1. Fertilization

5.1.1. General Considerations

Fertilization in general encompasses a sequence of three basic events: contact of a spermatozoon with an oocyte at the investments of an oocyte; fusion of the external membranes of the gametes; and finally, the actual mixing of the maternal and paternal genomes (i.e., syngamy) at the metaphase of the first mitotic division. Specifically, fertilization of a mammalian oocyte involves: (1) passage of a spermatozoon through the cellular layers of an oocyte, (2) attachment of the spermatozoon to the surface of the zona pellucida – a process that for some species is apparently mediated by species-specific receptors for spermatozoa (*Nicolson, Yanagimachi,* and *Yanagimachi,* 1975; *Oikawa, Yanagimachi,* and *Nicolson,* 1973), (3) passage through the zona pellucida, (4) fusion of the spermatozoal and oocyte plasma membranes (*Austin,* 1975), and (5) the actual penetration of the fertilizing spermatozoon into the oocyte with the resultant decondensation and dispersion of chromatin (*Gwatkin,* 1976, 1977). Fine structural aspects of each of these events have been studied by scanning and transmission electron microscopy, and the following references are recommended: *Austin* (1968), *Bedford* (1970), *Zamboni* (1971), *Gould* (1975), *Sugawara, Takeuchi,* and *Hafez* (1975), and *Gwatkin* (1977).

It was first clearly demonstrated by *Szollosi* and *Ris* (1961) that mammalian fertilization involves the fusion of the limiting membranes of the gametes, thus producing a single cell with a diploid complement of chromosomes and bounded by a membrane that, at least initially, is a composite or mosaic of the two gamete membranes (*Austin,* 1975). Additional correlates of the fertilization process are the so-called cortical and zona reactions which, either individually or in combination, prevent polyspermy – i.e., the penetration of an oocyte by accessory or supernumerary spermatozoa (see section 5.1.4). Both the fusion of the gametes and the block(s) to polyspermy involve cellular, molecular, and physiological modifications in the cortex of the oocyte and in the oocyte membrane (*Austin,* 1968; *Johnson,* 1975; *Johnson* and *Howe,* 1975; *Gwatkin,* 1976, 1977; *Solter,* 1977; *Yanagimachi,* 1977).

The events associated with the attainment of the fertilizable state of the male gamete have been described in Chapter 4. In the following section, some of the more pronounced events that occur in the female gamete prior to fertilization and that lead to the attainment of the fertilizable state are discussed.

5.1.2. The Oocyte During the Period Immediately Preceding Fertilization

Mammalian oocytes are ovulated with chromosomal maturation "arrested" at the second metaphase of meiosis. Typically, the first polar body, which had been extruded at the first meiotic division, is still present in the perivitelline space (Plate 56A) and is often connected to the oocyte by means of a cytoplasmic bridge (the midbody) containing microtubules (Plates 56B, 56C). The fine structure of the first polar body is quite similar to that of the oocyte insofar as it contains the same complement of cytoplasmic organelles (Plate 56B) (*Odor* and *Renninger,* 1960; *Zamboni* and *Mastroianni,* 1966; *Baca* and *Zamboni,* 1967; *Zamboni,* 1970, 1971). However, in contrast to the oocyte, the surface of the first polar body in the mouse is completely devoid of microvilli (Plate 56A). Major features of the period immediately preceding fertilization are the disruption of tight and gap junctions between coronal cell processes and the oolemma, and the retraction of these processes through the zona pellucida (Plate 15). Prior to ovulation, the continued contact of coronal cells and the oocyte appears to be essential for the normal growth and meiotic maturation of the mouse oocyte (*Eppig,* 1977; *Gilula, Epstein,* and *Beers,* 1978), as well as for the maturation of the pig (*McGaughey* and *Van Blerkom,* 1977) and rabbit oocyte (*Van Blerkom,* 1977; *Van Blerkom* and *McGaughey,* 1978a). With some variation among species, coronal cell processes are withdrawn from the surface of the oocyte either during the first or second meiotic metaphases (*Meyers, Young,* and *Dempsey,* 1936; *Dawson* and *Friedgood,* 1940; *Blan-*

dau, 1955; *Sotelo* and *Porter*, 1959; *Odor*, 1960; *Zamboni* and *Mastroianni* 1966; *Zamboni*, 1970). Although the actual mechanism of retraction likely involves the contractile proteins actin and myosin, which are located within the extensive microfilamentous bundles that occupy the processes (see Chapters 1 and 2), the precise nature of the stimulus that triggers retraction is unknown. However, as observed by transmission (*Nicosia, Wolf,* and *Inoue,* 1977) and scanning electron microscopy (*Nicosia, Wolf* and *Mastroianni,* 1978), a pronounced topographical reorganization of the oocyte plasma membrane occurs after the withdrawal of the coronal cell processes. For example, approximately 20% of the surface of an ovulated mouse oocyte, especially the region of the vitelline membrane that overlies the second meiotic spindle, has a relatively smooth appearance owing to the absence of microvilli (Plates 57 A, 58 A). By contrast, the remaining 80% of the plasma membrane of a mouse oocyte has a ruffled appearance and a dense population of microvilli (Plates 57 B, 58 B, 58 C). This view differs significantly from the situation that prevails in the mouse oocyte prior to the breakdown of the germinal vesicle (the initial event in meiotic maturation), in which the entire surface of the oocyte is covered by a dense population of uniformly distributed microvilli.

At the intracellular level, the formation of electron-dense granules from elements of the peripheral Golgi complexes (Plate 57 B) (*Szollosi,* 1967) and the distribution of these granules within the cortical cytoplasm are major features of the prefertilization period. In the mouse, numerous and prominent Golgi complexes (Plate 57 B) actively engage in the formation of cortical granules in the ovulated oocyte up to fertilization (Plate 57 B) (*Zamboni,* 1970). In comparison to a newly ovulated mouse oocyte, the number of cortical granules increases markedly during the immediate prefertilization period. Although variations in the spatial distribution of cortical granules have been reported (*Szollosi,* 1967; *Zamboni,* 1970), the studies of *Nicosia, Wolf,* and *Inoue* (1977) are the first to demonstrate the existence of "substantial gradients of cortical granules" in ovulated but unpenetrated mouse oocytes. Their observations revealed that approximately 20% of the total cortex homolateral to the second meiotic spindle is devoid of cortical granules, and, as mentioned above, the plasma membrane overlying the second meiotic spindle does not display microvilli. The remaining 80% of the cortex contains a morphologically heterogeneous population of granules and a plasma

membrane that exhibits numerous microvilli and a ruffled appearance. Most significantly, these authors noted that the fusion of the gametes rarely occurs in the region of the vitelline membrane devoid of cortical granules and microvilli. Immunofluorescent studies of unpenetrated mouse oocytes by *Johnson, Eager, Muggleton-Harris,* and *Grave* (1975) and *Eager, Johnson,* and *Thurley* (1976) indicate the presence of a mosaicism in the organization and distribution of conconavalin A receptor sites in the vitelline membrane (i.e., glycoproteins containing terminal mannose or glucose residues). Their studies demonstrated that the plasma membrane overlying the second meiotic spindle displayed a markedly reduced capacity to bind concanavalin A, possibly resulting from a parallel reduction in the density of microvilli in this region. Both morphological and immunofluorescent observations support the existence of a *polarity* in the cortex of the oocyte and in the overlying plasma membrane, which may preclude from certain areas the necessary juxtaposition of the gametes required for fusion (*Nicosia, Wolf,* and *Inoue,* 1977).

The question of when morphological and/or macromolecular polarities are established in the vitelline membrane is central to the elucidation of the role of polarities in the process of oocyte maturation and fertilization. The observations of *Stefanini, Oura,* and *Zamboni* (1969), as well as those of *Thompson, Moore-Smith,* and *Zamboni* (1974), indicate that the area of the cortical cytoplasm that overlies the second meiotic spindle in the mouse oocyte contains a filamentous/fibrillar layer. These authors suggested that the presence of this layer was involved in the abstriction of the second polar body. *Nicosia, Wolf,* and *Inoue* (1977) also described the existence of a filamentous layer and postulated that the filaments play an essential role in cytokinesis, possibly by increasing the rigidity of the gel strength of a specific cortical area. Because similar morphological characteristics are present in the rat oocyte prior to the extrusion of the first polar body (i.e., smooth surface, absence of cortical granules) (*Odor* and *Renninger,* 1960), it is conceivable that the origin of the polarity in the mammalian oocyte may be temporally and spatially related to the breakdown of the germinal vesicle and to the resumption of meiotic maturation (*Overstreet* and *Bedford,* 1974).

The presence of a gradient, polarity, or mosaicism in the structural organization of the cortex and the spatial distribution of macromolecules in the vitelline membrane are relatively new concepts as applied to mammalian oocytes and ova. Such polarities are common in the oocytes of amphibians and insects where the entrance and/or penetration of a spermatozoon is limited to a specific region of the plasma membrane. Further experimental investigations should provide clues relevant to (1) the origin of the polarities, (2) whether the spatial distribution of cortical granules affects the macromolecular organization and composition of the overlying plasma membrane (or vice versa), and (3) whether morphological and/or biochemical polarities are a common feature of mammalian oocytes and ova.

5.1.3. The Fusion of the Gametes

After traversing the zona pellucida, an acrosome-reacted mammalian spermatozoon reaches the vitelline membrane rather rapidly, although the duration of residence in the perivitelline space prior to contact varies among and within species. Several investigators have questioned the necessity of the acrosome reaction as an essential prerequisite for fusion (see, for example, *Thompson, Moore-Smith,* and *Zamboni,* 1974); however, it is apparent from the extensive studies of *Yanagimachi* and his collaborators (see *Yanagimachi,* 1977) that only acrosome-reacted spermatozoa are capable of undergoing fusion with the vitelline membrane. Detailed transmission electron microscopic examinations of the process of fusion have shown that, in contrast to fertilization in invertebrates, the inner acrosomal membrane of a mammalian spermatozoon *does not* fuse with the oocyte plasma membrane. Rather, fusion is usually accomplished between the vitelline membrane and the portion of the spermatozoal plasma membrane that covers the posterior region of the head (see Plate 47A, Chapter 4) (*Pikó* and *Tyler,* 1964; *Austin,* 1968; *Bedford,* 1970, 1972, 1974; *Yanagimachi* and *Noda,* 1970a, 1970b, 1970d; *Zamboni,* 1970, 1971, 1972; *Thompson, Moore-Smith,* and *Zamboni,* 1974; *Soupart* and *Strong,* 1974; *Anderson, Hoppe, Whitten,* and *Lee,* 1975; *Noda* and *Yanagimachi,* 1976; *Yanagimachi,* 1977). One possible exception to this scheme is that in some species the posterior region of the inner acrosomal membrane, where it joins the outer acrosomal membrane, may be involved in fusion. The mechanics of fusion appear to involve a two-step process: (1) fusion of the vitelline membrane with the portion of the spermatozoal plasma membrane extending from the posterior margin of the acrosome to the tip of the tail, and (2) engulfment of

the anterior two-thirds of the head within a cytoplasmic vacuole formed from the portions of the oocyte cytoplasm that surround the anterior region of the head (*Pikó,* 1969; *Stefanini, Oura,* and *Zamboni,* 1969; *Bedford,* 1970, 1972, 1974; *Yanagimachi* and *Noda,* 1970a, 1970b, 1970d; *Zamboni,* 1971, 1972; *Thompson, Moore-Smith,* and *Zamboni,* 1974). It appears likely that receptor molecules present on the surface of an oocyte are capable of interacting only with a specific population of spermatozoal macromolecules located in the posterior plasma membrane, and that this interaction provides for species-specific binding. Whether such macromolecules are modified either during residence of spermatozoa in the female reproductive tract or during passage through the cellular and noncellular investments of the oocyte is unknown.

Upon fusion of the gametes, a series of morphogenetic events occurs in the cortex of the egg cytoplasm that prevents additional spermatozoa (either in the vicinity of the zona pellucida or in the perivitelline space (Plate 59), depending upon the species) from undergoing fusion. These events are known as the cortical and zona reactions (see section 5.1.4). Following the fusion of the gametes, the spermatozoal nuclear membrane undergoes rapid disintegration, and the chromatin contained within the nucleus decondenses. Consequences of these processes are the activation of the egg, the resumption of meiosis, the abstriction of the second polar body, the formation of the male and female pronuclei, and syngamy. These aspects of mammalian fertilization are discussed in section 5.1.5.

5.1.4. The Cortical and Zona Reactions

The existence of cortical granules in mammalian oocytes and their behavior during fertilization were first described by *Austin* (1956) from light microscopic observations of golden hamster oocytes. *Austin* reported that after fertilization, most of the cortical granules disappeared from the peripheral cytoplasm and were presumed to have been discharged into the perivitelline space. More recent investigations have clearly shown that either fusion of the gametes (*Gwatkin, Rasmusson,* and *Williams,* 1976) or parthenogenetic activation of oocytes (*Van Blerkom* and *Runner,* 1976) is required for cortical granule release. The discharge of cortical granules and the release of the cortical contents into the perivitelline space appear to occur by a process of exocytosis. This process is known as the cortical reaction. In many species, the cortical reaction does not require contact between an oocyte and a spermatozoon since parthenogenetic activation (osmotic, temperature, electrical shock, or exposure to hyaluronidase, neuraminidase and lectins) will induce a cortical reaction (*Gwatkin, Rasmusson,* and *Williams,* 1976). It is generally considered that the cortical reaction results in the release of "factors" that (1) render the oocyte membrane impenetrable to accessory spermatozoa (Plate 59), and (2) alter the structure (*Baranska, Konwinski,* and *Kujawa,* 1975) and/or molecular composition of the zona pellucida (*Oikawa, Yanagimachi,* and *Nicolson,* 1973) possibly through the inactivation or removal of species-specific receptor sites for spermatozoa. This latter process is termed the zona reaction and is observed in many, but not all, fertilized mammalian eggs (*Austin* and *Braden,* 1956; *Austin,* 1961; *Barros* and *Yanagimachi,* 1971; *Gwatkin, Williams, Hartmann,* and *Kniazuk,* 1973; *Inoue* and *Wolf,* 1975; *Yanagimachi,* 1977).

Although the material contained within cortical granules is clearly associated with the fertilization process, its precise role and site(s) of action are still a matter of controversy. *Nicosia, Wolf,* and *Inoue* (1977) reported that some cortical granules in mouse oocytes are released prior to gamete fusion and have suggested that the premature release could aid in the initial attachment of a spermatozoon to the vitelline membrane by adding new material or by modifying cell surface receptors. On the other hand, the role of cortical granules in the block to polyspermy has been demonstrated in the ova of numerous mammals (*Gwatkin, Williams, Hartmanna,* and *Kniazuk,* 1973). In cases where spontaneous polyspermy is observed, such as in the rabbit, electron microscopic examinations show that cortical granules are retained in the egg, suggesting a defect in the cortical reaction (*Gulyas,* 1974a, 1974b).

The components of cortical granules that have been implicated in the block to polyspermy or in facilitating the initial attachment of the gametes include a trypsin-like protease (*Gwatkin, Williams, Hartmann,* and *Kniazuk,* 1973) as well as lectins, mucopolysaccharides, and glycoproteins (*Yanagimachi* and *Chang,* 1961; *Szollosi,* 1967; *Fléchon,* 1970; *Yanagimachi,* 1977). In the rat and mouse, the blocks to polyspermy involve changes in the physical and chemical composition of both the vitelline membrane and the zona pellucida. By contrast, the primary block to polyspermy in the hamster is restricted to the zona pellucida. In the rabbit, the primary block to poly-

spermy is at the level of the plasma membrane since numerous spermatozoa are present in the zona pellucida and perivitelline space after fertilization (Plate 59) (*Cooper* and *Bedford,* 1971; *Gorden, Fraser,* and *Dandekar,* 1975).

The hypothesis that cortical granules could have different functions in fertilization suggests that these granules may be present as a heterogeneous morphological or biochemical population and, further, that they may be differentially released into the perivitelline space after ovulation and at specific stages of the fertilization process. The electron microscopic observations of *Nicosia, Wolf,* and *Inoue* (1977) and *Nicosia, Wolf,* and *Mastroianni* (1978) described above offer tentative support for this hypothesis.

The accumulated evidence from numerous studies strongly suggests that in most mammalian oocytes, the cortical reaction, acting either at the level of the plasma membrane and/or the zona pellucida, is the primary mechanism preventing polyspermy. The following passage, derived from *Yanagimachi* (1977), summarizes the current interpretation of how the species specificity of the cortical reaction may function during fertilization:

"Regardless of whether the active components of cortical granules are enzymes or lectin-like substances, it is quite possible that these components modify the biological and biochemical characteristics of the zona material or the surface components of the egg plasma membrane by attacking or binding to these materials. If the active component is a trypsin-like enzyme, it would attack peptide chains in the glycopeptide complex, causing changes in the three-dimensional configuration of the complex as well as in the sequence of terminal saccharide residues of the complex. If it is a carbohydrate-splitting enzyme, it would attack carbohydrate moieties of the complex resulting in drastic changes in the sequence of the terminal saccharide residues. If it is a lectin-type substance, it would bind to the terminal saccharide chains. Once such changes occur in the zona or the surface component of the egg plasma membrane, the spermatozoa would no longer be able to recognize or bind to these egg components. The molecular configuration (e.g., the sequence of amino acids in the peptide chains and of the terminal saccharide residues of the carbohydrate moiety) of the zona pellucida and of the surface component of the egg plasma membrane must be genetically determined, thus the problems of species-specificity of the sperm-egg interactions (e.g., specificity of sperma-zona interactions) and the block to polyspermy may well be explained on "the same molecular basis or principle."

However, of the millions of spermatozoa present in the ejaculate, the number of spermatozoa located at the site of fertilization in some mammals is approximately equal to the number of ovulated oocytes (*Braden*, 1952; *Braden, Austin,* and *David,* 1954; *Austin,* 1968; *Bedford,* 1972). Yet, the organism must avail itself of a specific and certain mechanism by which polyspermy may be prevented, since the entrance of even a second spermatozoon would prove deleterious to normal development. Consequently, it would appear that in many mammals, the block(s) to polyspermy, although necessary, need not be as immediate as in invertebrates, in which the number of spermatozoa surrounding an oocyte at fertilization is significantly greater.

5.1.5. Postfusion Processes

The fusion of the gametes is accompanied by the zona and cortical reactions, the resumption of meiotic maturation and the so-called "activation" of the egg. Postfusion activation may be considered to encompass the initial cellular and molecular events associated with early embryogenesis (see section 5.2). The origin and nature of the activating stimulus (or stimuli) have traditionally been thought to involve agents present either within (*Bedford,* 1968, 1970, 1974; *Yanagimachi* and *Noda,* 1970c; *Brown* and *Hartree,* 1974) or on the surface of acrosome-reacted spermatozoa (*Morton,* 1975, 1977). However, as shown by *Yanagimachi* (1972) and *Yanagimachi, Yanagimachi,* and *Rogers* (1976), if spermatozoa do contain egg-activating factors, these factors are not species-specific, nor are they necessarily mandatory. In support of this observation is the production of apparently normal preimplantation mouse embryos (*Van Blerkom* and *Runner,* 1976; *Kaufman* and *Sachs,* 1976) and early fetuses (*Tarkowski, Witkowska,* and *Nowicka,* 1970; *Witkowska,* 1973a, 1973b; *Kaufman, Barton,* and *Surani,* 1977) from parthenogenetically activated oocytes (*Graham,* 1974). Although parthenogenetic embryos are at present unable to develop into normal mice, when chimeric embryos are formed by the aggregation of normal and parthenogenetically derived blastomeres, the resulting embryos do develop into apparently normal mice at birth (*Stevens, Varnum,* and *Eicher,* 1977).

From evidence currently available, it seems most likely that the activating stimulus involves alterations in the electrical potential of the egg membrane associated with changes in both the rate of influx and efflux of monovalent cations and the equilibrium between bound and free divalent cations (such as calcium). *Yanagimachi* (1977) and others have interpreted the process of activation as originating with the fusion of the gametes and a consequent *local* change in the permeability of the vitelline membrane to ions. Indeed, contemporary thought on activation suggests that intracellular calcium may be a *universal factor* that promotes the activation of egg metabolism and the associated molecular and cellular events of fertilization (*Mazia,* 1937; *Steinhardt, Epel, Carroll,* and *Yanagimachi,* 1974; *Wolf,* 1974). Whatever the actual nature and mechanism of activation may be, experimental findings strongly suggest that it is not species-specific, nor does it necessarily require the participation of a spermatozoon.

Shortly after the incorporation of a fertilizing spermatozoon into the oocyte cytoplasm, the nuclear membrane of the spermatozoon disintegrates (probably by means of a lytic process), and its chromatin begins to swell and decondense into filamentous strands which then diffuse into the cytoplasm (*Austin,* 1961; *Zamboni,* 1971; *Bedford,* 1972; *Longo,* 1973; *Thompson, Moore-Smith,* and *Zamboni,* 1974; *Anderson, Hoppe, Whitten,* and *Lee,* 1975). It is apparent from both biochemical and physiological studies that spermatozoal chromatin is both exceptionally stable and rich in disulfide bonds (*Bedford* and *Calvin,* 1974). Exposure of the chromatin to the hydrating conditions of the oocyte cytoplasm is insufficient to cause decondensation (*Calvin* and *Bedford,* 1971; *Bedford* and *Calvin,* 1974; *Mahi* and *Yanagimachi,* 1975; *Uehara* and *Yanagimachi,* 1976; *Usui* and *Yanagimachi,* 1976). What has been inferred from numerous observations of the early postfusion phase of fertilization is the presence in the oocyte cytoplasm of a factor or a set of factors (possibly enzymes that reduce disulfide bonds?) that are responsible for the decondensation of the chromatin. These postulated factors – termed "male pronucleus growth factor" by *Thibault* and *Gerard* (1973) and *Thibault* (1973, 1977) and "sperm-nucleus decondensing factors" by *Yanagimachi* (1977) – are thought to be synthesized in the preovulatory oocyte following the breakdown of the germinal vesicle (*Usui* and *Yanagimachi,* 1976; *Thibault,* 1977). At present, the existence of these factors is still only hypothetical. After decondensation, spermatozoal chromatin usually lies in a subcortical location, over which the plasma membrane, generally devoid of microvilli, protrudes markedly. Shortly after this state has been attained, a nuclear envelope develops to enclose the chromatin, thus forming the male pronucleus. The nuclear envelope develops by the union of vesicular membranes that have accumulated in the vicinity of the chromatin – a process identical to nuclear envelope formation in meiotic or mitotic cells (*Gwatkin,* 1976).

Concomitant with the fusion of the gamete plasma membranes is the resumption of meiotic maturation, culminating in the abstriction of the second polar body (Plate 58). Prior to fertilization, the second meiotic metaphase spindle is generally located parallel to the vitelline membrane. However, during the initial stages of fertilization, the equatorial plate rotates 90 degrees, such that one pole is adjacent to the surface of the egg while the other is located somewhat deeper within the cytoplasm (Plate 58 A). At telophase of the second meiotic division, the pole nearest the surface contains half of the chromosomes. The chromosomes, along with a small quantity of cytoplasm, are extruded from the oocyte as the second polar body (Plates 58 B, 58 C). In the mouse, as in other species, the second polar body remains attached to the fertilized egg for a protracted period of time, usually until at least the advanced pronuclear stage (*Stefanini, Oura,* and *Zamboni,* 1969). The more internal set of chromosomes quickly becomes surrounded by a double-membrane envelope, resulting in the formation of the female pronucleus. The mechanism of membranous enclosure of the chromosomes is similar to that described for the male pronucleus (Plate 57 C).

After the enclosure of the chromosomes, the two pronuclei enlarge significantly, move toward the central portion of the ovum, and come to lie in juxtaposition. Depending upon the species, as many as 30 or more nucleoli are present in the pronuclei (*Austin,* 1968). In the rabbit and human, each pronucleus appears to be nearly identical in size and structure and to contain a fibrous nucleoplasm in which fine-textured and granular aggregations are embedded (*Longo* and *Anderson,* 1969; *Zamboni,* 1971). The closely apposed pronuclei in the rabbit (but not the mouse) exhibit highly convoluted and extensive interdigitations of the nuclear membranes (*Longo* and *Anderson,* 1969), suggesting the possibility of internuclear communication (*Gondos* and *Bhiraleus,* 1970). In mammals, the association of the maternally and paternally derived genomes, a process termed "syngamy," occurs in the absence of the fusion of the pronuclear membranes (which is common to invertebrates, for example). During the growth and movement of the pronuclei, DNA synthesis is initiated, resulting in the duplication of the chromosomes. Chromosomal duplication in the rabbit egg occurs approximately 3 to 6 hours after fertilization (*Oprescu* and *Thibault,* 1965; *Szollosi,* 1966). The actual association of the maternal and paternal genomes is affected when the two pronuclear envelopes break down (the numerous nucleoli also disappear). The breakdown of the pronuclear envelopes and the condensation of the maternally and paternally derived chromosomes will also occur even if the pronuclei fail to migrate centrally (*Longo,* 1973).

After the breakdown of the pronuclear membranes, the two groups of chromosomes assume positions on the first cleavage spindle for the first mitotic division of the egg. The first mitotic division of the rabbit egg is rather unusual, insofar as during early telophase, individual chromosomes attain their own nuclear envelope. The individual, membrane-bound chromosomes of the rabbit are termed karyomeres (*Gulyas,* 1972). As observed by this investigator, chromosomes in the rabbit egg do not clump into one mass before reconstructing the nuclear envelope, as is common in other mammals (*Austin,* 1968; *Zamboni,* 1971). Instead, individual karyomeres fuse by means of pseudopod-like extensions or chromosomal bridges. Eventually, the fusion process becomes more extensive, and a single condensed nucleus, typical of the two-cell rabbit embryo, is formed. For a more detailed discussion of the fine structural events associated with syngamy and the formation of the two-cell embryo, the excellent study of *Longo* and *Anderson* (1969) is recommended. Their paper also contains additional and useful references relevant to classical as well as contemporary observations of syngamy and the first mitotic division.

Following the mixing of the maternal and paternal genomes and the first mitotic or cleavage division, the zygote enters into a phase of growth and development characterized by major changes in the organization of the cytoplasm and its components, as well as by changes in macromolecular syntheses. The period of postfertilization development, which includes cleavage and blastocyst formation, is termed the preimplantation period.

5.2. The Preimplantation Stages of Embryogenesis

The first formal description of a preimplantation rabbit conceptus was by *de Graaf* in 1672. Confirmation of the actual existence of the mammalian ovum and the first descriptive reports of the early stages of preimplantation development were published by *von Baer* (1827, 1828). Interest in this phase of embryogenesis continued through the 19th century, and certainly a contemporary researcher cannot but appreciate the abilities and insights of such early embryologists as *Coste* (1937), *Barry* (1838, 1839, 1840), and *Van Beneden* (1880), whose observations on developing embryos are generally valid to this day (see *Bodemar*, 1971, for a concise presentation of the history of mammalian embryology).

With the application of the electron microscope and elegant biochemical techniques to the study of early development, it has become possible to define more clearly the sequence of cellular and macromolecular steps that prepare the conceptus for implantation. Notwithstanding some species variation, numerous electron microscopic investigations have shown that preimplantation embryos undergo fundamentally similar patterns of subcellular differentiation.

In the same sense, analyses of preimplantation protein synthesis (*Brinster*, 1973; *Epstein* and *Smith*, 1974; *Van Blerkom* and *Manes*, 1974, 1977; *Van Blerkom* and *Brockway*, 1975; *Schultz* and *Tucker*, 1977; *Van Blerkom* and *McGaughey*, 1978b), RNA synthesis *(Graham*, 1973; *Manes*, 1975, 1977; *Schultz* and *Church*, 1975; *Van Blerkom* and *Manes*, 1977), metabolism and enzymatic activities (*Biggers* and *Stern*, 1973), and physiology (*Biggers* and *Borland*, 1976) have also demonstrated fundamental similarities among the few different species studied. Although the following sections describe morphological aspects of preimplantation development, relevant observations derived from biochemical and molecular biological studies are introduced to provide a more meaningful discussion of the cellular processes that characterize the preimplantation period. The scanning electron microscope has provided basic information relevant to topographical and architectural alterations that accompany early mammalian embryogenesis (see section 5.4).

However, most of the major developmental processes occur at the intracellular level, and, therefore, the emphasis of the following sections is placed on observations obtained by transmission electron microscopy.

5.2.1. The Cytoplasm of the Fertilized Mammalian Ovum and Developing Embryo

5.2.1.1. Lattice-like structures

One of the most prominent characteristics of the cytoplasm of the newly fertilized eggs of some mammals is a dense population of "lattice-like" structures (Plates 60A, 60B). Transmission electron microscopic studies of rodents (mouse, rat, hamster) have shown that the cytoplasm of maturing oocytes and cleaving embryos contains highly ordered arrays of fibrous-like strands which exhibit cross-striations resulting in a lattice-like appearance (*Enders* and *Schlafke*, 1965; *Szollosi*, 1965a; *Weakley*, 1966, 1967, 1968; *Schlafke* and *Enders*, 1967; *Calarco* and *Brown*, 1969; *Zamboni*, 1970, 1971; *Van Blerkom* and *Runner*, 1976). Differences in the morphological organization of these structures have been observed between the mouse, hamster, and rat (*Schlafke* and *Enders*, 1967; *Weakley*, 1968). In the mouse, numerous cross-linked chains form a typical lattice-like configuration (Plates 60A, 60B) (*Calarco* and *Brown*, 1969). Only single chains exist in the rat (*Schlafke* and *Enders*, 1967), while in the hamster, double chains are found which resemble ladder-like structures. By contrast, the oocytes and preimplantation embryos of the rabbit (*Gulyas*, 1971a; *Van Blerkom, Manes*, and *Daniel*, 1973), baboon (*Panigel, Kraemer, Kalter, Smith*, and *Heberling*, 1975), sheep (*Calarco* and *McLaren*, 1976), and human (*Baker* and *Franchi*, 1967; *Soupart* and *Strong*, 1974, 1975) do not contain these elements.

Both the precise nature and the developmental significance of the lattice-like aggregates are not completely understood. *Manzanek* (1965) suggested that they might represent linear arrays of ribosomes, and in the same sense, *Zamboni* (1970) has proposed that they are a form of polyribosomes. Histochemical and biochemical studies indicate that the lattices are composed of RNA and protein (*Schlafke* and *Enders*, 1967; *Weakley*, 1967; *Burkholder, Comings*, and *Okada*, 1971). Enzymatic and electron microscopic observations of whole-mount preparations of mouse oocytes provide evidence supporting a ribosomal nature (*Burkholder, Comings*, and *Okada*, 1971). The findings of these investigators suggest that the lattices are a storage form of inactive, maternal ribosomes, which after fertilization, are used in the support of protein synthesis until the embryo produces its own functional ribosomes. In support of this interpretation are the reports of *Bachvarova (* 1975) and *Bachvarova* and *De Leon* (1977) that mouse oocytes contain a store of inactive ribosomes which are mobilized immediately after fertilization and used in protein synthesis. Although the concept of a storage form of ribosomes is attractive, the definitive nature of the lattice-like arrays requires their isolation, a more precise biochemical investigation (for example, do they contain ribosomal proteins?), and a determination of whether the components of these arrays are active in protein synthesis. If these structures do indeed represent a storage form of ribosomes that are activated after fertilization, then it must be determined why in some species they persist into the blastocyst stages when preimplantation embryos are normally synthesizing new ribosomes (Plate 60B)(*Schlafke* and *Enders,* 1967; *Grant, Nilsson* and *Bergstrom,* 1977; and section 5.3.1). In any event, that structures of this nature are encountered in and necessary for only some mammals may be a potentially significant observation, insofar as it may indicate fundamental alternatives in the molecular biology of early mammalian embryogenesis.

5.2.1.2. Virus-like particles

A-type. One of the most intriguing and perplexing observations to come from the electron microscopic examination of preimplantation embryos is the demonstration of cytoplasmic particles that closely resemble A-type RNA viruses. A-type particles (presumed) have been observed in guinea pig oogonia (*Anderson* and *Jeppesen,* 1972) and in guinea pig (*Enders* and *Schlafke,* 1965) and rabbit trophoblast (Plates 61A, 61B) (*Manes,* 1973). At present, the best documented examples of the occurrence and distribution of these particles are in the oocytes and early embryos of the mouse (Plates 60A, 60C, 60D). *Calarco* and *Szollosi* (1973) reported that A-type particles were present in oocytes prior to germinal vesicle breakdown, disappeared during meiotic maturation, but reappeared in the cytoplasm of two-cell embryos (Plate 60A). The particles are a prominent feature of the cytoplasm between the two- and eight-cell stages (Plates 60C, 60D), but typically they are present in greatly reduced numbers after the eight-cell stage (they are rarely encountered at the blastocyst stage) (*Calarco* and *Brown,* 1969; *Calarco* and *Szollosi,* 1973; *Chase* and *Pikó,* 1973; *Calarco,* 1975a; *Van Blerkom* and *Runner,* 1976). That A-type particles also appear at the two-cell stage in parthenogenetically activated embryos demonstrates that the paternal genome is not required for their formation (*Biczysko, Solter, Graham,* and *Koprowski,* 1974; *Van Blerkom* and *Runner,* 1976). The accumulated electron microscopic evidence of RNA viral-like particles supports a vertical and perhaps maternal mode of transmission — i.e., the genomic information required for particle formation is an integral part of the genetic composition of the mouse, and therefore the formation of the particles can be considered to be the expression of essentially normal cellular genes.

Although at least three size classes of putative A-type particles have been described for preimplantation mouse embryos (*Chase* and *Pikó,* 1973), the form most abundant in both the mouse and rabbit measures approximately 100 nm in diameter and is composed of an inner sphere of approximately 39 nm with an electron lucent center and an outer sphere approximately 67 nm in diameter (Plate 60A) (*Calarco* and *Szollosi,* 1973). In the mouse, A-type particles are usually located within expanded cisternae of the smooth-surfaced endoplasmic reticulum (intracisternal) (Plates 60A, 61A, 61B). In the rabbit, the particles are encountered in clusters, free within the cytoplasm (intercisternal), and, usually, adjacent to an electron-dense, somewhat granular mass (Plate 61A). *Calarco* (1975a) suggests that the putative A-type particles are formed in regions of the embryonic cytoplasm containing a fibrous material which is adjacent to elements of the smooth-surfaced endoplasmic reticulum. The particles are presumed to form from this material and then bud into the cisternae. Other investigators have indicated a nuclear participation in their formation and elaboration (Plate 60D) (*Van Blerkom* and *Runner,* 1976).

The disappearance of A-type particles in oocytes, and their reappearance at the two-cell stage, coincide with the cessation and resumption of ribosomal RNA synthesis in the mouse (see section 5.3.1, and *Hillman* and *Tasca,* 1969; *Knowland* and *Graham,* 1972; *Calarco* and *Szollosi,* 1973). In addition, A-type particles fail to appear at the two-cell stage in mouse embryos that have been exposed to levels of actinomycin D known to inhibit ribosomal RNA synthesis (*Calarco,* 1975a). These observations suggest a correspondence between particle formation and the synthesis of ribosomal RNA. However, if such a correspondence does exist, it must be explained why the particles are usually not encountered in the cytoplasm of late cleaving/early blastocyst-stage mouse embryos — stages associated with active ribosomal RNA synthesis. In spite of the observation that A-type particles in some mammalian oocytes and early embryos are morphologically similar to A-type particles in other cells (*Sarkar, Moore,* and *Nowinski,* 1972), there is no definitive biochemical evidence to indicate (1) that the particles in oocytes and embryos are viruses, (2) whether they have properties, other than a similar morphology, in common with known A-type RNA viruses, and (3) whether they have any functional significance in early mammalian embryogenesis.

C-type. In contrast to A-type particles, C-type particles are more readily identifiable because they are shed from a cell by "budding" off the plasma membrane and accumulate in intercellular spaces. C-type particles are RNA viruses (oncornaviruses) and in mammals are commonly encountered in the tissues of leukemic animals. C-type particles have a characteristic morphology, i.e., a membrane-bound particle approximately 100 nm in diameter, possessing a distinct, electron-dense core or nucleoid. Several investigators have reported observing particles morphologically similar to C-type particles in preimplantation mouse (*Chase* and *Pikó,* 1973; *Biczysko, Solter, Graham,* and *Koprowski,* 1974), rabbit (*Manes,* 1973) (Plates 61 C, 61 D), and baboon embryos (*Kalter, Panigel, Kraemer, et al.,* 1974; *Panigel, Kraemer, Kalter, Smith,* and *Heberling,* 1975), as well as in the placental tissues of several primates (including man) (*Kalter, Helmke, Panigel, et. al.,* 1973; *Kalter, Helmke, Heberling, et al.,* 1973; *Schidlovsky* and *Ahmed,* 1973; *Vernon, McMahon,* and *Hackett,* 1974). In comparison to A-type particles (especially in the mouse), C-type particles are encountered infrequently in preimplantation stages, and according to some reports, the possibility exists that they may have been mistaken for other structures (such as microvilli) or extracellular debris.

As noted by *Jaenisch* and *Berns* (1977), the detection of particles in preimplantation embryos and postimplantation tissues that are morphologically similar to C-type viruses is a strong indication both of the expression of viral genes contained within the embryonic genome and of vertical transmission of the genes between generations. Possibly the best evidence to date demonstrating the expression of C-type viral genes in preimplantation embryos is the study by *Pikó* (1977). He reported that immunocytochemical staining with antibodies raised against the major core protein (p 30) of C-type murine leukemic viruses detected an antigen in both the germinal vesicles of mouse oocytes and the nuclei of preimplantation embryos up to the morula stage of development. *Pikó* (1977) has interpreted these observations as providing a direct indication of the existence of an oocyte

and embryonic protein that is (1) immunologically similar to the viral p 30 protein, and (2) a product of a cellular gene that has a normal function in embryonic development. However, the recent study of *Strand, August,* and *Jaenisch* (1977) suggests that while variations in gene expression of the p 30 antigen are evident during early mouse embryogenesis, and may even be influenced by the state of differentiation of the cell in which it is expressed, such expression does not necessarily support the concept that C-type viral genes, even if integrated into the mouse genome, play an important role in mamalian embryogenesis.

An alternative hypothesis, favored by *Huebner, Kelloff, Sarma, et al.* (1970) and by *Todaro* and *Huebner* (1972), suggests that while the genes for endogenous RNA viruses may be important determinants of leukemia in the adult, they may also be important determinants of development during early embryogenesis (the viral oncogene hypothesis), perhaps by altering the composition of the glycoprotein component of the surface of certain embryonic cells. At present, assigning any function to either A- or C-type particles in early development is highly speculative and the ultimate determination of the viral nature of the particles, as well as an elucidation of function, await additional analysis and more definitive experiments.

5.2.1.3. The nuage

The germinal plasm of some insects and amphibians contains accumulations of dense, fibrous bodies which have been directly implicated in the determination and continuity of the germ line (the germ plasm hypothesis) (*Illmensee* and *Mahowald,* 1974; and reviewed by *Hegner,* 1914; *Wilson,* 1925; *Eddy,* 1975; *Smith* and *Williams,* 1975). In *Drosophila* these bodies are specifically localized to the posterior pole plasm of eggs and embryos, and, consequently, they have been termed "polar granules" (*Hegner,* 1914; *Wilson,* 1925; *Mahowald,* 1962, 1968, 1971). In amphibians, structures similar to the polar granules of insects have been observed in the vegetal hemisphere of eggs amidst groups of mitochondria (*Kessel,* 1969; *Mahowald* and *Hennen,* 1971; *Williams* and *Smith,* 1971) and in the putative germinal plasm of developing embryos (*Smith* and *Williams,* 1975); these particular elements are known as germinal granules. Morphologically similar structures, referred to as the "nuage" (after *André* and *Rouiller,* 1957, who described this material in insects), have been encountered in primordial germ cells, oogonia, oocytes, spermatogonia, spermatocytes, spermatids, and preimplantation embryos of numerous mammals (*Hadek* and *Swift*, 1962; *Szollosi*, 1965b; *Eddy*, 1970, 1974, 1975; *Fawcett, Eddy,* and *Phillips*, 1970; *Gulyas*, 1972; *Motta* and *Van Blerkom,* 1974). The nuage of mammalian germ cells and embryos exists either (1) as fine granular and/or filamentous material situated in the interstices of mitochondrial clusters (Plate 62A), or (2) as discrete, dense, fibrous masses that lack a surrounding membrane and are freely distributed in the cytoplasm (Plates 62B, 62C, 62D). Frequently, cytoplasmic nuage assumes a form characteristic of nucleoli, and therefore this material has been referred to in the literature as nucleolar-like bodies, pseudonucleoli, or nucleolar granules (Plate 62B). In male germ cells, the nuage forms discrete and consistent structures known as chromatoid bodies (*Eddy*, 1970, 1975).

The origin and chemical nature of the nuage in mammals is still a controversial issue. It is not clear whether it is composed of RNA and/or protein or whether it has a nuclear or nonnuclear origin (*Eddy*, 1975). Although the nuage in developing mammalian gametes and embryos has a fine structure similar to the polar granules of insects and the germinal granules of amphibians, morphological criteria alone are insufficient to equate the nuage to a putative germ plasm determinant. If the nuage is responsible for the determination of germ cells in mammals, it should be traceable from the gametes through embryogenesis to the primordial germ cells (*Eddy*, 1974). Whereas the morphological observations mentioned above are consistent with the concept that the nuage may be involved in the segregation of germ cell plasma in mammals, it is essential to demonstrate experimentally whether nuage is even required for germ cell formation. As noted by *Eddy* (1974), the germ plasm hypothesis is appealing because of its relative simplicity and its seemingly broad applicability; however, it has yet to be rigorously tested in mammals.

5.2.1.4. Crystalloids and crystals

A characteristic feature of the cytoplasm of cleaving mouse embryos is the presence of masses of a crystalline material that are not membrane-bound (Plate 60B). *Calarco* and *Brown* (1969) reported the presence of crystalline material in two-cell embryos, but noted that the number and size of the crystalloids increased markedly during cleavage and blastocyst formation (especially after the four-cell stage). The crystalloids also appear in parthenogenetic mouse embryos during cleavage, but become less obvious at the blastocyst stages (*Van Blerkom* and *Runner*, 1976). At the late morula stage, crystalloids are usually associated with cisternae of rough-surfaced endoplasmic reticulum (Plate 60B). The association of crystalloids with the rough-surfaced endoplasmic reticulum and their histochemical properties (*Moskalewski, Sawicki, Gabara,* and *Koprowski*, 1971) indicate that they are proteinaceous in nature. *Calarco* and *Brown* (1969) suggested that these structures may represent a protein product synthesized and stored during the preimplantation period for use during later development. The actual origin and function of these crystalloids in early mouse development are, however, obscure.

One of the more striking examples of crystalline inclusions in preimplantation embryos is observed in the rabbit blastocyst between 4 and 4.5 days following fertilization. Originally described by *Van Beneden* in 1880, the blastocyst crystals are visible in the light microscope (Plate 63A), are membrane-bound, have a discernible repeating regularity (Plates 63B, 64), and are frequently in excess of 5μm in length (*Hadek* and *Swift*, 1960; *Tyndale-Biscoe*, 1965; *Van Blerkom, Manes,* and *Daniel*, 1973; *Hoffman, Davies,* and *Long*, 1975). Electron and light microscopic studies demonstrate that (1) crystals are not encountered in the cells of the inner cell mass (Plate 64A), (2) they fail to appear in rabbit embryos cultured *in vitro* from the one-cell to the expanded blastocyst stage, and (3) they are not present in the trophoblast cells of embryos reared *in vivo* for approximately 3.5 to 4 days postfertilization (i.e., up to about 12 hours before crystals would normally appear) followed by culture *in vitro* for 48 hours (*Van Blerkom, Manes,* and *Daniel*, 1973; *Van Blerkom* and *O'Farrell*, unpublished observations). Collectively, the evidence obtained from the above experiments indicates either that the material that composes the crystals is of maternal origin or that the continued presence of the embryo in the reproductive tract is essential for their formation. Alternatively, *Daniel* and *Kennedy* (1978) have recently indicated that rabbit trophoblast crystals may arise from pre-existing structures present in the newly fertilized egg.

An examination of the fine structural changes in the rabbit endometrium during the progestational phase of preimplantation development shows that crystalline bodies of the same ultrastructural appearance as those observed in blastocysts become evident in the glandular epithelium on day 3.5 following fertilization (*Davies* and *Hoffman,* 1975). Although the relative number of crystals observed in endometrial cells is significantly lower than the number observed in the trophoblast, the appearance of endometrial crystals is temporally correlated with rising or maximal secretion of progesterone. Strengthening the relationship between circulating levels of progesterone and the appearance of endometrial and embryonic crystals are the reports that (1) crystals appear in the endometrium on day 3.5 of pseudopregnancy, and (2) their appearance in the endometrium of ovariectomized rabbits can be induced by the administration of progesterone, but not of estradiol-17β (*Nako, Meyer,* and *Noda,* 1971; *Hoffman, Davies,* and *Long,* 1975). Although the possibility that the crystals arise from pre-existing structures or material in the rabbit embryo cannot be excluded at present (*Daniel* and *Kennedy,* 1978), it seems equally probable that they are derived, at least in part, from material secreted by the progestational uterus. The observation that crystals are not encountered in the cells of the inner cell mass of an expanded rabbit blastocyst could be explained by the fact that these cells, as contrasted with the cells of the trophoblast, are not directly exposed to the uterine milieu.

Histochemical and cytochemical studies indicate that the rabbit crystals are proteinaceous in nature (*Hadek* and *Swift,* 1960; *Nako, Meyer,* and *Noda,* 1971; *Hoffman, Davies,* and *Long,* 1975). Our own studies of isolated crystals demonstrate that they are composed of several proteins (possibly glycoproteins) with individual molecular weights of approximately 50,000 daltons (*Van Blerkom* and *O'Farrell,* unpublished results). The appearance of rabbit trophoblast crystals at both the light and electron microscopic levels bears a striking similarity to crystals of catalase. However, both enzymatic studies of whole rabbit blastocysts and of isolated crystals failed to show catalase activity (*Van Blerkom,* unpublished).

The developmental significance of blastocyst crystals in the rabbit is unknown. However, the evidence presented above suggests that certain proteins may be acquired and stored by trophoblast cells in a highly concentrated form, possibly for some developmental process that occurs subsequent to implantation, when the crystals progressively disappear. The relative abundance of crystals in rabbit blastocysts, coupled with the ability to isolate them, should provide the basis for future experimentation and a more precise biochemical definition of their composition, as well as a more precise determination of their origin and function in embryonic development.

5.3. The Differentiation of the Major Cellular Organelles

5.3.1. Nucleolar Differentiation, Ribosomal RNA Synthesis, and the Formation of Polysomes

One of the strongest correlations of cell structure and function observed during the preimplantation stage is the progressive differentiation (maturation) of nucleoli and the concomitant onset of ribosomal RNA synthesis and ribosome formation. Two morphological patterns of nucleolar maturation have been described for preimplantation embryos, and, in the following discussion, these patterns are represented by the mouse and the rabbit. In both species, nucleoli in newly fertilized eggs are spherical, osmophilic bodies composed of a dense, fibrillar matrix (Plate 65). As embryos progress through cleavage, a granular element is detected and nucleoli undergo a sequential process of reticulation (vacuolation). In rabbit (*Van Blerkom, Manes,* and *Daniel,* 1973) and sheep embryos (*Calarco* and *McLaren,* 1976), reticulation appears to encompass the entire structure (Plates 66 A, 66 B). In the mouse (*Calarco* and *Brown,* 1969; *Hillman* and *Tasca,* 1969; *Van Blerkom* and *Runner,* 1976) and rat (*Schuchner,* 1970; *Szollosi,* 1971), by contrast, reticulation begins at the periphery of the organelle (Plates 66 C, 66 D, 73 A). By the late morula stage in the mouse and at the early blastocyst stage in the rabbit, nucleoli have fully differentiated into hypertrophic and extensively reticulated structures in which granular and fibrillar materials are distributed along highly anastomosing networks of nucleolonemas (Plate 67). The morphological detail of differentiated nucleoli in preimplantation mammalian embryos is typical of nucleoli in cells that are actively engaged in the synthesis of ribosomal RNA (*Busch* and *Smetana,* 1970; *Miller* and *Gonzales,* 1976).

A characteristic feature of mammalian embryos during the early stages of cleavage is a scarcity of cytoplasmic ribosomes and polysomes (Plates 68 A, 68 B) (*Calarco* and *Brown,* 1969; *Van Blerkom, Manes,* and *Daniel,* 1973). However, coincident with the appearance of a granular element and the progressive reticulation of the nucleolar matrix, ribosomes and polysomes become more abundant. With the eventual formation of mature nucleoli, ribosomes and polysomes densely populate the cytoplasm (Plates 68 C, 68 D). Although the stage of development at which polysomes become a prominent feature of the cytoplasm varies among species, it is always associated with the appearance of a granular element in the nucleoli and with the onset of detectable ribosomal RNA synthesis *(Hillman* and *Tasca,* 1969; *Graham,* 1973; *Manes,* 1975; *Schultz* and *Church,* 1975; *Van Blerkom* and *Manes,* 1977). For example, ribosomal RNA synthesis is first detected in the mouse embryo between the two- and four-cell stage and increases progressively thereafter (*Schultz* and *Church,* 1975). By contrast, autoradiographic (*Karp, Manes,* and *Hahn,* 1973) and radioisotopic labeling techniques (*Manes,* 1971, 1977) have failed to detect significant levels of ribosomal RNA synthesis in the rabbit during most of cleavage, although transfer RNA and heterogeneous RNA are produced (*Schultz,* 1973, 1975; *Manes,* 1977; *Schultz* and *Tucker,* 1977); significant ribosomal RNA synthesis is apparently "delayed" in the rabbit until the initial stages of blastocyst formation (*Manes,* 1971). Of course, these observations do not preclude the possibility that some ribosomal RNA synthesis is occurring prior to the blastocyst stage, but at a level below the sensitivity of the techniques currently available for its detection. Thus, even though the timing of nucleolar differentiation, ribosomal RNA synthesis, and the production of ribosomes differs considerably among species, the correlation of nucleolar fine structure and macromolecular syntheses is a common feature of all mammalian embryos studied to date (see reviews by *Manes,* 1975; *Schultz* and *Church,* 1975; *Schultz* and *Tucker,* 1977; *Van Blerkom* and *Manes,* 1977).

5.3.2. Mitochondrial Differentiation

As with most of the fine structural changes that accompany early mammalian embryogenesis, the differentiation of maternally derived mitochondria follows a sequential pattern as well (*Calarco* and *Brown,* 1969; *Schlafke* and *Enders,* 1967; *Hillman* and *Tasca,* 1969; *Anderson, Condon,* and *Sharp,* 1971; *Stern, Biggers,* and *Anderson,* 1971; *McReynolds* and *Hadek,* 1972; *Van Blerkom, Manes,* and *Daniel,* 1973; *Van Blerkom* and *Runner,* 1976). In the newly fertilized oocyte and cleaving rabbit embryo, mitochondria are relatively small, electron-dense, spheroidal organelles (Plate 68 A). These mitochondria contain few cristae; the cristae are present in the periphery of the organelle and are oriented in a circular fashion, parallel to the outer mitochondrial membranes. At approximately the 16-cell stage, many mitochondria appear somewhat more elongated than previously and contain cristae that partially penetrate

the mitochondrial matrix which is still electron-dense (Plate 68 B). Coincident with the transition from the morula to the blastocyst stage, mitochondria undergo a striking morphodynamic alteration, such that by the mid-blastocyst stage, two morphological forms are apparent. Mitochondria of the first type are elongated, contain a matrix of low electron density, and possess numerous lamellar cristae arranged in a stack-like manner, perpendicular to the long axis of the organelle (Plate 63 B). This morphological form is encountered predominantly in trophoblast cells. The second type of mitochondrion observed in rabbit blastocysts is also elongated, but contains a matrix of moderate-to-high electron density and distended cristae surrounding electron translucent spaces (vacuolated mitochondria) (Plate 68 C). Mitochondria with this appearance predominate in the cells of the inner cell mass (*Van Blerkom, Manes,* and *Daniel,* 1973).

Throughout preimplantation development, mouse (*Calarco* and *Brown,* 1969) and rat embryos (*Schlafke* and *Enders,* 1967) contain vacuolated mitochondria (*Wischnitzer,* 1967) – an appearance that results from the presence of distended cristae enclosed within a dense mitochondrial matrix (Plate 68 D). During the initial cleavage divisions, the number of cristae per mitochondrion is relatively small. However, during the latter phases of cleavage (after the eight-cell stage), mitochondria become more elongated and the number of cristae increases significantly. Our own observations of serially sectioned embryos demonstrate the presence of the same two morphological forms in mouse blastocysts as was described above for the rabbit, although mitochondria with lamellar cristae were encountered less frequently than in the rabbit. In this regard, *Grant, Nilsson,* and *Bergstrom* (1977) have recently reported the existence of two similar morphological forms of mitochondria in hamster blastocysts, but no special distribution between inner cell mass and trophoblast cells was noted.

Changes in mitochondrial fine structure during the preimplantation stages are correlated rather closely with shifts in the metabolism and metabolic patterns of the embryos (*Biggers* and *Stern,* 1973; *Biggers* and *Borland,* 1976). The metabolic rate of preimplantation embryos, based on the rate of oxygen consumption, has been measured for the rabbit *(Fridhandler,* 1957) and the mouse (*Mills* and *Brinster,* 1967). Measurements of the metabolic activity demonstrate that an increase in the utilization of oxygen occurs at the mid-morula stage in the mouse and at the late morula-early blastocyst stage in the rabbit. Therefore, it seems reasonable to suggest that alterations in the fine structure of embryonic mitochondria, such as a decrease in the relative density of the matrix and a concomitant increase in the number of cristae per mitochondrion, may reflect the level of metabolic activity in which the organelle is engaged (*Anderson, Condon,* and *Sharp,* 1971; *Biggers* and *Stern,* 1973; *Van Blerkom, Manes,* and *Daniel,* 1973; *Biggers* and *Borland,* 1976). The developmental and functional significance of the two apparent morphological forms of mitochondria in mouse, hamster, and rabbit blastocysts is unknown at present.

5.3.3. Membrane-Granule Complexes and the Appearance of Rough-Surfaced Endoplasmic Reticulum

An intriguing morphodynamic process occurring at the mid-morula stage of preimplantation development in the rabbit is the appearance of "membrane-granule complexes" (Plates 69, 70) (*Van Blerkom, Manes,* and *Daniel,* 1973). To date, these whorl-like accumulations of membranes and electron-dense granules have only been reported for the rabbit embryo and are encountered in the cytoplasm of all blastomeres between the mid-morula and very early blastocyst stages. Although the complexes are generally located in a juxtanuclear position, they are also observed in proximity to the plasma membrane (Plates 69, 70). Within these whorl-like structures are observed (1) regularly repeating, electron-dense bodies (granules) of a fairly uniform size and shape (Plates 69B, 70), and (2) larger osmiophilic bodies in the more central portions of the complexes (Plate 70B). Serial reconstruction of thin-sectioned rabbit morulae demonstrates that the complexes are spheroidal in nature and occupy a significant proportion of the cytoplasm (Plate 69B). The precise origin of the membrane complexes is unknown; however, serial section reconstructions have provided some useful insights (*Van Blerkom, Manes,* and *Daniel,* 1973). For example, three characteristic features of the cytoplasm of cleaving rabbit embryos are the presence of distended profiles of the smooth-surfaced endoplasmic reticulum, Golgi saccules (Plate 71A), and vesicles containing a "flocculent" material (Plates 70A, 71B, 71C, 71D) (flocculent vesicles; *Gulyas,* 1971b). Evidence from serial reconstructions indicates that (1) some of the flocculent vesicles may arise from the nucleus by "budding" off the outer nuclear membrane (Plates 70A, 71C) (flocculent vesicles have also been suggested to give rise to annulate lamellae in rabbit embryos; *Gulyas,* 1971b), (2) the membranes of these particular flocculent vesicles are often continuous with the membranes of the complexes located in perinuclear positions (Plate 70A), and (3) the central portion of many flocculent vesicles contains electron-dense bodies, similar in size and shape to the osmiophilic structures located within the whirls of membranes (Plates 71B, 71C, 71D). Collectively, morphological observations suggest that the membrane-granule complexes may be derived in part from both Golgi elements and flocculent vesicles (*Van Blerkom, Manes,* and *Daniel,* 1973). In addition, cisternae containing an amorphous material of moderate electron density appear to arise from the periphery of the complexes (Plates 70B) and to migrate to the more cortical regions ot the cytoplasmic. Alternatively, they may be formed by the fragmentation of flocculent vesicles (Plate 71D).

Although the functional and developmental significance of the complexes is obscure, the presence of electron-dense bodies within the flocculent vesicles, and the apparent structural continuity between the complexes and the membranes of Golgi saccules and flocculent vesicles, raise the possibility that the complexes are involved in the metabolism and/or processing of the material contained within the flocculent vesicles. An alternative possibility is that the complexes are involved in the generation, assembly, and/or storage of membrane precursors or specialized proteins (perhaps an embryonic component of the blastocyst crystals, as suggested by *Daniel* and *Kennedy,* 1978) required for the rapid expansion and development of the blastocyst (*Van Blerkom, Manes,* and *Daniel,* 1973). The determination of the origin and function of these complexes is one of the major unanswered questions relevant to the morphophysiological development of the preimplantation rabbit embryo.

Cisternae of the rough-surfaced endoplasmic reticulum are rarely observed in the cytoplasm of cleaving mouse and rabbit embryos (*Calarco* and *Brown,* 1969; *Hillman* and *Tasca,* 1969; *Van Blerkom, Manes,* and *Daniel,* 1973). However, after the eight-cell stage in the mouse and between the late morula and early blastocyst stage in the rabbit, cisternae of rough-surfaced endoplasmic reticulum become prominent features of the cytoplasm (Plate 68D) and, in fact, are often observed to be continuous with the outer nuclear membrane (the site of formation) (Plate 68E). By contrast, numerous profiles of the smooth-surfaced endoplasmic reticulum are encountered during cleavage, but become increasingly rare as the embryo progresses through the latter phases of cleavage, especially during the blastocyst stages. The timing of the formation of the rough-surfaced endoplasmic reticulum in both the mouse and the rabbit coincides with the differentiation of nucleoli into organelles active in ribosomal RNA synthesis and the appearance of a dense population of cytoplasmic ribosomes and polysomes (*Calarco* and *Brown*, 1969; *Hillman* and *Tasca*, 1969; *Van Blerkom*, *Manes*, and *Daniel*, 1973). In other cellular systems, the primary function of the rough-surfaced endoplasmic reticulum is the synthesis of proteins destined for secretion (*Fawcett*, 1966). If in mammalian embryos this organelle has a comparable function, then proteins synthesized on these membranes could conceivably be involved in defining the macromolecular composition of the cell surface and/or in the formation of junctional complexes (see following section).

5.3.4. The Formation of Intercellular Junctions and Junctional Complexes and Correlated Changes in the Cell Surface and Cortical Cytoplasm: Factors Involved in the Early Differentiation of the Embryo

The development of intercellular junctions and junctional complexes in preimplantation embryos is a critical event in early embryogenesis. These junctions not only provide the structural basis for fluid accumulation and expansion during the blastocyst stage, but also may play a direct role during early cleavage in the establishment and maintenance of cell position within the embryo – functions related to the concept that cell position during cleavage may determine the differentiative fate of individual blastomeres (*Tarkowski* and *Wroblewska*, 1967; *Hillman*, *Sherman*, and *Graham*, 1972; *Herbert* and *Graham*, 1974). The development, distribution, and characterization of junctional complexes during preimplantation development discussed in the following sections have been derived from studies of rabbit (*Van Blerkom*, *Manes*, and *Daniel*, 1973; *Hastings* and *Enders*, 1975), rat (*Schlafke* and *Enders*, 1967; *Enders*, 1971), and mouse embryos (especially the excellent papers of *Ducibella* and his collaborators: *Ducibella*, *Albertini*, *Anderson*, and *Biggers*, 1975; *Ducibella* and *Anderson*, 1975; *Ducibella*, *Ukena*, *Karnovsky*, and *Anderson*, 1977; *Ducibella*, 1977).

5.3.4.1. Early cleavage

Typically, early cleaving embryos lack well-developed intercellular junctions, and, for the most part, large intercellular spaces separate blastomeres (Plates 72 A, 73 A). Where cells are in contact, membranous associations are maintained by means of microvilli and other cellular projections: gap junctions are occasionally observed in areas of intercellular contact (Plate 72 B). As observed by scanning electron microscopy, microvilli appear to be uniformly distributed on the cell surface (Plate 72 C) (*Calarco* and *Epstein*, 1973).

5.3.4.2. The morula stage and compaction

To appreciate fully the developmental significance of the formation of intercellular junctions during the morula stage, it is necessary to discuss briefly the requirements for cavitation and the resultant establishment of the blastocyst. As noted by *Ducibella* (1977) and others, the formation of the blastocyst is dependent on the development of an outer layer of a fluid-transporting epithelia (the trophoblast or trophectoderm). These cells are interconnected by means of zonular tight junctions which maintain a permeability seal between the interior portion of the embryo, containing the inner cell mass, and the milieu of the reproductive tract (*Enders*, 1971). The outer layer of trophoblast could not develop if the blastomeres retained the spherical shape that characterized them during cleavage, because round blastomeres can only establish focal (spot) or macular intercellular contacts which are incapable of functioning as effective seals. Consequently, in order to provide the necessary structural conditions required for cavitation and expansion during blastocyst formation, a continuous, close apposition of the lateral margins of the blastomeres must take place. Such a process involves changes in cell shape and the migration of portions of blastomeres over one another. This morphogenetic process is known as compaction, and the time during cleavage when this phenomenon begins and is completed varies among species.

At the early eight-cell stage in the mouse, individual blastomeres are spherical and possess distinct cell outlines (Plate 73 B). On the surface of the blastomeres, numerous microvilli appear to be uniformly distributed (a situation that prevailed from the two-cell stage), although the density of these microvilli decreases where cells are in close apposition (Plate

76 A). Likewise, from the one-cell to the early eight-cell stage, mitochondria, ribosomes, crystalloids, and fibrillar elements appear to be uniformly distributed throughout the cytoplasm (Plate 65 B). However, at some point during the eight-cell stage, the mouse embryo undergoes compaction. Compaction in the mouse (and in some other species) is accompanied by (1) the maximization of intercellular contact between apposing cells (Plate 74), (2) the loss of distinct cell borders when viewed by scanning electron microscopy (Plate 74), (3) a flattening of the blastomeres located on the "outside" of the embryo, (4) the appearance of focal tight junctions which then progressively develop into zonular tight junctions, and (5) changes in the distribution of subcellular components in the cortical cytoplasm. As early as 1894, *Assheton* observed the close packing of blastomeres which occurred in the eight-cell rabbit embryo, but it is not until the mid-morula (16- to 32-cell stage) that rabbit embryos display a typical compacted appearance (Plate 75) (*Lewis* and *Gregory*, 1939).

At the cellular level, compaction is associated with the formation of tight and gap junctions, the former appearing initially at the apical margins of apposing cells as focal (spot) tight junctions. Coincident with macroscopic parameters discussed above, continuous tight junctions (zonular tight junctions) develop circumferentially around the cellular borders. Considering the fact that in other cellular systems the establishment and maintenance of tight junctions require calcium (*McNutt* and *Weinstein*, 1973), it is not surprising that the compaction of the mouse embryo *in vitro* is calcium-dependent (*Wales*, 1970; *Ducibella* and *Anderson*, 1975). Consequently, because compaction involves changes in the nature of cell-cell adhesion, it is not unlikely that calcium-dependent adhesive forces are involved (*Ducibella* and *Anderson*, 1975). In addition, compaction could be associated with a change in the rigidity of the plasma membrane. In support of this hypothesis are the observations that compaction in the mouse is accompanied by a redistribution of cortical cytoskeletal elements in which microtubules become oriented parallel to the apposed membranes of blastomeres (*Ducibella, Ukena, Karnovsky,* and *Anderson,* 1977). These investigators have suggested that the orientation and localization of microtubules may be involved in the stabilization of the plasma membrane and in the determination of cell shape. Likewise, the finding that mouse morulae may be "decompacted" by exposure of the embryos *in vitro* to cytochalasin B (the effect is reversible upon removal of the drug) indicates the

participation of microfilaments in this process. An additional observation of *Ducibella, Ukena, Karnovsky,* and *Anderson* (1977) which may be of fundamental importance in the development and differentiation of the early mouse embryo is that mitochondria undergo a redistribution during compaction such that, in contrast to the situation that prevailed prior to compaction, mitochondria become localized in the cortical cytoplasm. This observation suggests that the energy requirements of the transport systems of epithelial cells (trophoblast or trophectoderm) may be provided by locally high concentrations of adenosine triphosphate (ATP) produced by cortical mitochondria relokated during compaction (*Ducibella, Ukena, Karnovsky,* and *Anderson,* 1977). Whether a similar redistribution occurs during compaction in other species remains to be determined.

As has been demonstrated by *Calarco* and *Epstein* (1973) and *Ducibella, Ukena, Karnovsky,* and *Anderson* (1977), compaction in the mouse involves a redistribution of microvilli on the surface of the blastomeres. Whereas prior to compaction microvilli are uniformly distributed (Plates 72 C, 73 B), the onset of compaction is accompanied by the restriction of microvilli to the apical surface and to the basal zone of intercellular contact, with the region between these localizations being relatively devoid of microvilli (Plate 76) (*Ducibella,* 1977). This redistribution suggests that basal microvilli play a role in approximating the plasma membranes of adjacent blastomeres and that a zone of microvilli moves apically in the intercellular furrow during compaction (*Ducibella,* 1977). Further strengthening the hypothesis that basal microvilli function in providing the necessary juxtaposition of plasma membranes required for the formation of zonular tight junctions is the observation of *Calarco* and *Epstein* (1973) that, prior to compaction, thin cellular projections and microvilli extend from one cell to another, possibly binding the cells more tightly together.

Thus, the calcium-dependent formation of zonular tight junctions between apposing blastomeres in early cleaving embryos provides not only an effective seal required for the accumulation of fluid within the interior of the embryo (the blastocoelic or blastocyst cavity) (Plate 77), but also creates a permeability barrier between the milieu of the reproductive tract and the internal portion of the morulae and, later, of the blastocyst. A barrier of this type may be of fundamental importance in the establishment of a microenvironment within the interior of the morula (the location of the presumptive inner cell mass). Such a microenvironment has been shown to exist in the early mouse embryo (*Borland,* 1977) and, it has been suggested that the creation of a microenvironment could be an important factor in the determination of molecular and cellular aspects of the inner cell mass (*McMahon,* 1974; *Ducibella,* 1977). The presence of gap junctions in compacted morulae (as well as in pre-compaction embryos) represents a potentially significant finding, because, as discussed in Chapters 1 and 2, these low resistance junctions are currently thought to mediate and coordinate intercellular communication (see recent review of *Griepp* and *Revel,* 1977). Therefore, if gap junctions function in a similar manner in early embryos, then a possible coordination between macromolecular syntheses and developmental fate could be established early in embryogenesis.

Taken together, available evidence indicates that compaction is a critical event in early mammalian embryogenesis since (1) cell surface and cortical changes during compaction are the first observable manifestation of polarity in *embryogenesis,* (2) cell surface and intercellular changes associated with the development of the trophoblast may occur during cleavage in the mouse, as well as in other species, long before a definitive trophoblast or trophectoderm appears at the blastocyst stage, and (3) the formation of zonular tight junctions effectively differentiates outside cells from inside cells for the first time during preimplantation development.

5.3.4.3. The blastocyst stage

The first "macroscopically" observable divergence of embryonic cells into two phenotypically distinct populations occurs at the blastocyst stage with the formation of the inner cell mass and the trophoblast. These two cell layers differ not only morphologically (*Van Blerkom, Manes,* and *Daniel,* 1973; *Ducibella, Albertini, Anderson,* and *Biggers,* 1975) but also immunologically (*Solter* and *Knowles,* 1975; *Wiley* and *Calarco,* 1975; *Muggleton-Harris* and *Johnson,* 1976). In the mouse (the only species examined in this respect at present), they also differ in their sensitivity to ^{3}H-thymidine (*Snow,* 1973) and in their qualitative pattern of protein synthesis (*Van Blerkom, Barton,* and *Johnson,* 1976; *Van Blerkom,* 1977; *Handyside* and *Barton,* 1977). Of equal importance is the finding that these two types of cells are committed to certain differentiative pathways and therefore exhibit limited developmental capacities and fates when manipu-

lated experimentally (*Gardner* and *Johnson,* 1972, 1973, 1975; *Gardner* and *Papaioannou,* 1975; *Gardner* and *Rossant,* 1976; *Snow, Aitken,* and *Ansell,* 1976; *Surani* and *Barton,* 1977). In general, embryonic structures are derived from inner cell mass, whereas extraembryonic structures (placental membranes) are trophoblast derivatives. Clearly, that the cells of these two layers are associated by means of different types of intercellular junctions is insufficient evidence to explain their very different molecular and developmental properties; however, an understanding of the nature and distribution of intercellular junctions in blastocyst-stage embryos provides a structural framework upon which the above-mentioned characteristics of inner cell mass and trophoblast cells may be viewed.

The blastocyst stage of many preimplantation embryos is accompanied by (1) the accumulation of fluid within the interior of the embryo, resulting in the formation of a blastocyst cavity or blastocoel (cavitation) and the characteristic swelling or expansion of the embryo, and (2), in some species, the active transport of ions (*Borland, Biggers,* and *Lechene,* 1977) and the selective uptake of proteins into the blastocoel (*Hastings* and *Enders,* 1974). Exceptions to this basic pattern of early embryonic development include the human and the armadillo, in which the period immediately preceding implantation does not involve significant embryonic expansion (*Enders,* 1971). In those species in which expansion does occur, however, extremes of size attained by blastocysts prior to implantation range from approximately 190 μm in diameter in the mouse to nearly 6 mm in diameter in the rabbit to over *one meter* in length in the pig. In most species, the swelling of the blastocoel results in the extreme attenuation of the outer layer of trophoblast cells, a marked decrease in the amount of membrane available for intercellular contact, the thinning of the zona pellucida, and the development in some species of significant hydrostatic forces within the blastocyst (Plate 78). Consequently, special structural requirements are placed upon trophoblast cells in order to continue and maintain the expanded state (although some embryos, such as the mouse, alternatively expand and collapse during the blastocyst stage). When examined by transmission electron microscopy, the existence of intercellular junctional complexes between adjacent trophoblast cells provides the ultrastructural basis for the maintenance of the expanded state (see *Schlafke* and *Enders,* 1967; *Enders, 1971; Van Blerkom, Manes,* and *Daniel,* 1973; *Ducibella, Albertini, Anderson,* and *Biggers,*

1975). Coincident with the cavitation of the rabbit embryo between 2.5 and 3.0 days following fertilization is the appearance of junctional complexes composed of zonular tight junctions, gap junctions, and forming or primitive desmosomes at the lateral borders of trophoblast cells (Plate 79 A). Forming desmosomes are recognized by the presence of an accumulation of electron-dense material in the space between apposing cells and by the presence of cytoplasmic filaments oriented parallel to the plasma membrane on either side of the dense, intercellular material (Plate 79 A). Zonular tight junctions at the apical borders of adjacent cells again serve to maintain a permeability barrier between the "inside" and the "outside" of the embryo – an absolute prerequisite for expansion (*Enders*, 1971). With the continued swelling of the rabbit blastocyst, elaborate desmosomes, typically in a tandem arrangement, appear between trophoblast cells on day 4 or 4.5 following fertilization (also shown in Plate 80 B). Associated with these desmosomes are tonofilaments and other subplasmalemmal filaments, some of which are arranged parallel to the plasma membrane (Plate 79 B), while others, organized into bundles, originate from deep within the cytoplasm and converge at the desmosome (Plates 79 C, 80). Elaborate desmosomes are only observed between trophoblast cells, where they presumably provide the necessary cell-cell anchoring (*Farquar* and *Palade*, 1963) required to maintain the structural continuity and integrity of the embryo by enabling the cells of the trophoblast to withstand the internal hydrostatic pressures generated during expansion. The gap junction component of the junctional complex could function in the coordination of cellular activity, such as the timing of mitosis. The basic sequence of events involved in junction formation described for the rabbit (see also *Hastings* and *Enders*, 1975) occurs in many other species whose embryonic development has been studied by transmission electron microscopy. These mammals include the rat (*Schlafke* and *Enders*, 1967; *Enders*, 1971), mouse (*Arguello* and *Martinez-Palomo*, 1975; *Ducibella, Albertini, Anderson,* and *Biggers,* 1975), baboon (*Fléchon, Panigel, Kraemer, Kalter,* and *Hafez*, 1976), and sheep (*Calarco* and *McLaren,* 1976). The formation of intercellular junctions in the trophoblast and the presence of tandem desmosomes again underscore the close association of cell structure and function during preimplantation development.

In comparison to trophoblast cells, cells of the inner cell mass are rounded and are not interconnected by means of junctional complexes (Plate 78 B). Neither zonular tight junctions nor elaborate tandem desmosomes are usually observed between inner cell mass cells; presumably, adhesion of these cells to one another is achieved primarily by means of gap junctions (Plate 77) *(Van Blerkom, Manes,* and *Daniel,* 1973; *Calarco,* 1975 b; *Hastings* and *Enders,* 1975; *Ducibella, Albertini, Anderson,* and *Biggers,* 1975). Tight junctions (focal) and small desmosomes have been observed between cells of the inner cell mass, but in comparison to the trophoblast, these intercellular membrane specializations are rather infrequent (*Enders,* 1971).

Intercellular contact between cells of the inner cell mass and trophoblast appears to involve primarily gap and adhering junctions which serve to bring these two cell types into close, local contact. In the mouse, trophoblast cells elaborate cellular projections which occupy spaces between cells of the inner cell mass and appear to both fix the inner cell mass at one pole of the embryo and to isolate it partially from the blastocoel (*Ducibella, Albertini, Anderson,* and *Biggers,* 1975). That inner cell mass cells do not become significantly attenuated during expansion and are not interconnected by elaborate, tandem desmosomes indicates that they may not have the same internal structural requirements as trophoblast cells. The absence of extensive junctional complexes in the inner cell mass is a cellular manifestation of a developmental program which differs from that displayed by the trophoblast. Because gap junctions are currently thought to act as mediators of intercellular communication by functioning in metabolic cooperation and ionic coupling between metabolically different cells (*Gilula, Reeves,* and *Steinbach,* 1972; *Bennett,* 1973; *McNutt* and *Weinstein,* 1973; *Gilula,* 1974), it is not unlikely that gap junctions between the various cells of a blastocyst act in a similar manner (*Ducibella,* 1977). In support of this hypothesis are experimental studies indicating the necessity of intercellular communication between cells of the inner cell mass, and between cells of the inner cell mass and trophoblast, for the normal development and differentiation of the blastocyst stage embryo (*Gardner,* 1975; *Lin,* 1969). Although the fine structural basis for intercellular communication exists in the blastocyst, the chemical nature of the signals that are presumably transferred between cells is unknown.

5.4. Surface Characteristics of the Blastocyst and the Formation of the Endoderm

The zona pellucida of the rabbit blastocyst is retained during the initial stages of implantation, while in the mouse and rat, the blastocyst completely "hatches" through the zona pellucida prior to implantation, the zona then being discarded. *Boving* (1972) has noted that the application of the term "zona pellucida" to the rabbit blastocyst is probably inaccurate. Instead, he considers that the noncellular envelope of the embryo is a trilaminar structure composed of an oolemma, mucolemma, and a gliolemma, which originates in the ovary, oviduct, and uterus, respectively. However, for descriptive purposes, the term "zona pellucida" is used here. As observed by scanning electron microscopy, the surface of the zona in the expanded rabbit blastocyst has a textured appearance and is covered by random accumulations of an amorphous material (Plate 81). The external surface of rabbit trophoblast (facing the perivitelline space) exhibits a dense population of elongated microvilli (Plate 81 B). This is in contrast to the internal surface of the trophoblast (facing the blastocoelic cavity) which displays a reduced number of microvilli (Plates 78 B, 80, 82 A). Such a polarity in the expression of microvilli may be a manifestation of different functions of the external and internal trophoblast membranes, or possibly a morphophysiological response to different microenvironments. In the hatched mouse blastocyst, the external surface of the trophectoderm displays a morphology ranging from smooth to microvillous. The presence of numerous microvilli in mouse blastocysts is usually associated with mitotic cells. The characteristic features of the external surface of preimplantation mouse blastocysts are smooth intercellular ridges averaging 0.6μm in width (Plate 82 B) (*Bergstrom*, 1972; *Calarco*, 1975 b) and tripartite Y-shaped structures that extend across junctional ridges (Plate 83 A) (*Calarco*, 1975 b). Similar structures are not observed in rabbit blastocysts, and it is not known at present whether the Y-shaped structures are surface specializations for adherence and communication or merely an accumulation of intercellular material that protrudes above the junctional complex (*Calarco*, 1975 b).

In comparison to other species, the mouse blastocyst does not expand significantly (Plate 82 B). The marked expansion of the rabbit blastocyst, by contrast, results in the development of extremely flattened cells (Plate 83 B). In both the rabbit and the mouse, the cells that compose the inner cell mass are relatively devoid of microvilli and have observable cell borders and cytoplasmic bulges which correspond to the region of the cytoplasm occupied by the nucleus (Plate 84 A).

Within the inner cell mass, two populations of cells become apparent – endodermal and ectodermal cells. Endoderm cells are generally the first of the inner cell mass to show cytological evidence of preimplantation differentiation (*Enders*, 1971). In most mammals, endoderm cells arise by delamination from the inner cell mass and eventually form a layer of cells that lines the interior surfaces of the inner cell mass and trophoblast.

Although the origin of the endoderm is clearly of fundamental importance in early mammalian development, only the morphogenetic events involved in its formation in the rabbit are considered. As observed by transmission electron microscopy, endodermal cells appear to arise from the inner cell mass as an extremely thin epithelium in which cells are interconnected by means of junctional complexes containing small desmosomes. As observed by scanning electron microscopy, associations between apposing cells are also maintained by means of overlapping cytoplasmic projections (Plate 84 C). Frequently, individual cells are found that are in the process of migrating over the inner surface of the trophoblast (*Enders*, 1971; *Enders, Given,* and *Schlafke,* 1978). In the scanning electron microscope, the extremely flattened and laminar nature of the parietal endoderm is quite apparent (Plate 84 B). Endodermal cells are generally devoid of microvilli, and the area of the cytoplasm containing the nucleus is usually not apparent from the cell surface (Plate 84 B). One of the more interesting features of the migration of the parietal endoderm is the appearance of elongated cellular projections that traverse fairly extensive intercellular gaps to connect with similar processes elaborated by other endodermal cells (Plates 84 B, 85 B). The surface of the trophoblast (trophectoderm) that faces the endoderm is also devoid of microvilli (Plates 84 B, 85). Occasionally, rounded cells displaying microvilli, numerous blebs, and an extensive network of filapodia are observed (Plate 85 A). These particular endodermal cells may be in mitosis, and the cellular evaginations probably function to anchor the cell to the trophoblast.

References

Andersen, H. K., and *Jeppesen, T.* (1972). Virus-like particles in guinea pig oogonia and oocytes. J. Nat. Cancer Inst. 49, 1403–1410.

Anderson, E., Condon, W., and *Sharp, D.* (1971). A study of oogenesis and early embryogenesis in the rabbit, *Oryctolagus cuniculus,* with special reference to the structural changes of mitochondria. J. Morphol. *130,* 67–92.

Anderson, E., Hoppe, P. C., Whitten, W. K., and *Lee, G. S.* (1975). *In vitro* fertilization and early embryogenesis: A cytological analysis. J. Ultrastruct. Res. *50,* 231–252.

André, J., and *Rouiller, C.,* (1957). L'ultrastructure de la membrane nucléaire des ovocytes de l'araignée *(Tegenaria domestica Clark).* In: Proceedings of the European Conference on Electron Microscopy (Stockholm, 1956), pp. 162–164. Academic Press, New York.

Arguello, C., and *Martinez-Palomo, A.* (1975). Freeze fracture morphology of gap junctions in the trophoblast of the mouse embryo. J. Ultrastr. Res. *53,* 271–283.

Assheton, R. (1894). A re-investigation into the early stages of the development of the rabbit. Quart. J. Micro. Sci. *37,* 113–164.

Austin, C. R. (1956). Cortical granules in hamster eggs. Exp. Cell Res. 10, 533–540.

Austin, C. R. (1961). The Mammalian Egg. Charles C. Thomas, Springfield, Ill.

Austin, C. R. (1968). Ultrastructure of Fertilization. Holt, Rinehart and Winston, New York.

Austin, C. R. (1975). Membrane fusion events in fertilization. J. Reprod. Fert. *44,* 155–166.

Austin, C. R., and *Braden, A. W. H.* (1956). Early reactions of the rodent egg to spermatozoon penetration. J. Exp. Biology *33,* 358–365.

Baca, M., and *Zamboni, L.* (1967). The fine structure of human follicular oocytes. J. Ultrastruct. Res. *19,* 354–381.

Bachvarova, R. (1975). Mouse egg ribosomes. J. Cell Biology *67* (suppl.), 14 a.

Bachvarova, R., and *De Leon, V.* (1977). Stored and polysomal ribosomes of mouse ova. Devel. Biol. *58,* 248–254.

Baker, T. G., and *Franchi, L. L.* (1967). The fine structure of oogonia and oocytes in human ovaries. J. Cell Sci. *2,* 213–224.

Baranska, W., Konwinski, M., and *Kujawa, M.* (1975). Fine structure of the zona pellucida of unfertilized egg cells and embryos. J. Exp. Zool. *192,* 193–202.

Barros, C., and *Yanagimachi, R.* (1971). Induction of zona reaction in golden hamster eggs by cortical granule material. Nature (London) *233,* 268–269.

Barry, M. (1838). Researches in embryology: First series. Phil. Trans. Roy. Soc. London, Pt. 1, 301.

Barry, M. (1839). Researches in embryology: Second series. Phil. Trans. Roy. Soc. London, Pt. 2, 307.

Barry, M. (1840). Researches in embryology: Third series; A contribution to the physiology of cells. Phil. Trans. Roy. Soc. London, Pt. 2, 529.

Bedford, J. M. (1968). Ultrastructural changes in the sperm head during fertilization in the rabbit. Am. J. Anat. *123,* 329–358.

Bedford, J. M. (1970). Sperm capacitation and fertilization in mammals. Biol. Reprod., suppl. 2, 128–158.

Bedford, J. M. (1972). Sperm transport, capacitation and fertilization. In: Reproductive Biology, pp. 338–392 (Balin, H., and Glasser, S., eds.). Excerpta Medica, Amsterdam.

Bedford, J. M. (1974). Mechanisms involved in penetration of spermatozoa through the vestments of the mammalian egg. In: Physiology and Genetics of Reproduction, Pt. B, pp. 55–68 (Coutinho, E. M., and Fuchs, F., eds.). Plenum Press, New York.

Bedford, J. M., and *Calvin, H. I.* (1974). The occurrence and possible functional significance of -s-s-crosslinks in sperm heads, with particular reference to eutherian mammals. J. Exp. Zool. *188,* 137–156.

Bennett, M. V. L. (1973). Function of electrotonic junctions in embryonic and adult tissues. Fed. Proc. *32,* 65–75.

Bergstrom, S. (1972). Delay of blastocyst implantation in the mouse by ovariectomy or lactation: A scanning electron microscope study. Fertil. Steril. *23,* 548–561.

Biczysko, W., Solter, D., Graham, C., and *Koprowski, H.* (1974). Synthesis of endogenous type-A virus particles in parthenogenetically stimulated mouse eggs. J. Nat. Cancer Inst. *52,* 483–489.

Biggers, J. D., and *Borland, R. M.* (1976). Physiological aspects of growth and development of the preimplantation mammalian embryo. Ann. Rev. of Physiol. *38,* 95–119.

Biggers, J. D., and *Stern, S.* (1973). Metabolism of the preimplantation mammalian embryo. Adv. Reprod. Physiol. *6,* 1–59.

Blandau, R. J. (1955). Ovulation in the living albino rat. Fert. Steril. *6,* 391–404.

Bodemar, C. W. (1971). The biology of the blastocyst in historical perspective. In: The Biology of the Blastocyst (Blandau, R. J., ed.). University of Chicago Press, Chicago and London.

Borland, R. M. (1977). Transport processes in the mammalian blastocyst. In: Development in Mammals, Vol. 1 (Johnson, M. H., ed.). North Holland Publishing Co., Amsterdam and New York.

Borland, R. M., Biggers, J. D., and *Lechene, C.* (1977). Studies on the composition and formation of mouse blastocoele fluid using electron probe microanalysis. Devel. Biol. *55,* 1–8.

Boving, B. G. (1972). Spacing and orientation of blastocyst in utero. In: Biology of Mammalian Fertilization and Implantation (Moghissi, K. S., and Hafez, E. S. E., eds.). Charles C. Thomas, Springfield, Ill.

Braden, A. W. H. (1952). Properties of the membranes of rat and rabbit eggs. Australian J. Scientific Research *5,* 460–471.

Braden, A. W. H., Austin, C. R., and *David, H. A.* (1954). The reaction of the zona pellucida to sperm penetration. Austral. J. Biol. Sci. *7,* 391–409.

Brinster, R. L. (1973). Protein synthesis and enzyme constitution of the preimplantation mammalian embryo. In: The Regulation of Mammalian Reproduction, pp. 302–334 (Segal, S., Crozier, R., Corfman, P. A., and Condliffe, P. G., eds.). Charles C. Thomas, Springfield, Ill.

Brown, C. R., and *Hartree, E. F.* (1974). Distribution of a trypsin-like proteinase in the ram spermatozoon. J. Reprod. Fert. *36*, 195–202.

Burkholder, G. D., Comings, D. E., and Okada, T. A. (1971). A storage form of ribosomes in mouse oocytes. Exp. Cell Res. *69*, 361–371.

Busch, H., and Smetana, K. (1970). The Nucleolus, pp. 59–105, 169–211. Academic Press, New York.

Calarco, P. G. (1975 a). Intracisternal A particle formation and inhibition in preimplantation mouse embryos. Biol. Reprod. *12*, 448–454.

Calarco, P. G. (1975 b). Cleavage (mouse). In: Scanning Electron Microscopic Atlas of Mammalian Reproduction (Hafez, E. S. E., ed.). Springer-Verlag, Berlin.

Calarco, P. G., and Brown, E. H. (1969). An ultrastructural and cytological study of preimplantation development of the mouse. J. Exp. Zool. *171*, 253–283.

Calarco, P. G., and Epstein, C. J. (1973). Cell surface changes during preimplantation development in the mouse. Devel. Biol. *32*, 208–213.

Calarco, P. G., and McLaren, A. (1976). Ultrastructural observations of preimplantation stages of the sheep. J. Embryol. Exp. Morph. *36*, 609–622.

Calarco, P. G., and Szollosi, D. (1973). Intracisternal A particles in ova and preimplantation stages of the mouse. Nature New Biol. (London) *243*, 91–93.

Calvin, H. I., and Bedford, J. M. (1971). Formation of disulphide bonds in the nucleus and accessory structures of mammalian spermatozoa during maturation in the epididymis. J. Reprod. Fert., suppl. 13, 65–75.

Chase, D. G., and Pikó, L. (1973). Expression of A- and C-type particles in early mouse embryos. J. Nat. Cancer Inst. *51*, 1971–1975.

Cooper, G. W., and Bedford, J. M. (1971). Charge density change in the vitelline surface following fertilization of the rabbit egg. J. Reprod. Fertil. *25*, 431–436.

Coste, J. J. M. C. V. (1837). Embryogénie Comparée. Paris.

Daniel, J. C., and Kennedy, J. R. (1978). Crystalline inclusion bodies in rabbit embryos. J. Embryol. Exp. Morph. *44*, 31–43.

Davies, J., and Hoffman, L. H. (1975). Studies on the progestational endometrium of the rabbit. II. Electron microscopy, day 0 to day 13 of gonadotrophin induced pseudopregnancy. Am. J. Anat. *142*, 335–366.

Dawson, A. B., and Friedgood, H. B. (1940). The time and sequence of preovulatory changes in the cat ovary after mating or mechanical stimulation of the cervix uteri. Anat. Rec. *76*, 411–424.

Ducibella, T. (1977). Surface changes in the developing trophoblast cell. In: Development in Mammals, Vol. 1 (Johnson, M. H., ed.). North Holland Publishing Co., Amsterdam and New York.

Ducibella, T., Albertini, D. F., Anderson, A., and Biggers, J. D. (1975). The preimplantation mammalian embryo: Characterization of intercellular junctions and their appearance during development. Devel. Biol. *45*, 231–250.

Ducibella, T., and Anderson, E. (1975). Cell shape and membrane changes in the eight-cell mouse embryo: Prerequisite for morphogenesis of the blastocyst. Devel. Biol. *47*, 45–58.

Ducibella, T., Ukena, T., Karnovsky, M., and Anderson, E. (1977). Changes in cell surface and cytoplasmic organization during early embryogenesis in the preimplantation mouse embryo. J. Cell Biol. *74*, 153–167.

Eager, D. D., Johnson, M. H., and Thurley, K. W. (1976). Ultrastructural studies on the surface membrane of the mouse egg. J. Cell. Sci. *22*, 345–353.

Eddy, E. M. (1970). Cytochemical observations on the chromatoid body of male germ cells. Biol. Reprod. *2*, 114–128.

Eddy, E. M. (1974). Fine structural observations on the form and distribution of nuage in germ cells of the rat. Anat. Record *178*, 731–758.

Eddy, E. M. (1975). Germ plasm and the differentiation of the germ line. In: International Review of Cytology, Vol. 43 (Bourne, G. H., Danielli, J. F., and Jeon, K. W., eds.). Academic Press, New York.

Enders, A. C. (1971). The fine structure of the blastocyst. I: The Biology of the Blastocyst, pp. 71–94. University of Chicago Press, Chicago.

Enders, A. C., and Schlafke, S. J. (1965). The fine structure of the blastocyst: Some comparative studies. In: Preimplantation Stages of Pregnancy, pp. 29–54 (Wolstenholme, G. E., and O'Conner, M., eds.) Little, Brown, and Co., Boston.

Enders, A. C., Given, R. L., and Schlafke, S. (1978). Differentiation and migration of endoderm in the rat and mouse at implantation. Anat. Rec. *190*, 65–78.

Eppig, J. J. (1977). Mouse oocyte development in vitro with various culture systems. Devel. Biol. *60*, 371–388.

Epstein, C. J., and Smith, S. A. (1974). Electrophoretic analysis of proteins synthesized by preimplantation mouse embryos. Devel. Biol. *40*, 233–244.

Farquar, M. G., and Palade, G. E. (1963). Junctional complexes in various epithelia. J. Cell Biol. *26*, 263–291.

Fawcett, D. W. (1966). The Cell, An Atlas of Fine Structure. W. B. Saunders, Philadelphia.

Fawcett, D. W., Eddy, E. M., and Phillips, D. M. (1970). Observations on the fine structure and relationships of the chromatoid body in mammalian spermatogenesis. Biol. Reprod. *2*, 129–153.

Fléchon, J. E. (1970). Nature glycoprotéique des granules corticaux de l'oeuf de lapin. Mise en évidence par l'utilisation comparée de techniques cytochimiques ultrastructurales. J. de Microscopie *9*, 221–242.

Fléchon, J.-E., Panigel, M., Kraemer, D. C., Kalter, S. S., and Hafez, E. S. F. (1976). Surface ultrastructure of preimplantation baboon embryos. Anat. Embryol. *149*, 289–295.

Fridhandler, L. (1957). Developmental changes in the respiratory activity of rabbit ova. Exp. Cell Res. *13*, 132–139.

Gardner, R. L. (1975). Analysis of determination and differentiation in the early mammalian embryo using intra- and inter-specific chimaeras. In: The Developmental Biology of Reproduction, pp. 207–238 (Markert, C. L., and Papaconstantinous, I., eds.), Academic Press, New York.

Gardner, R. L., and Johnson, M. H. (1972). An investigation of inner cell mass and trophoblast tissues following their isolation from the mouse blastocyst. J. Embryol. exp. Morph. *28*, 279–312.

Gardner, R. L., and Johnson, M. H. (1973). Investigation of early mammalian development using interspecific chimaeras between rat and mouse. Nature New Biol. (London) *246*, 86–89.

Gardner, R. L., and Johnson, M. H. (1975). Investigation of

cellular interaction and deployment in the early mammalian embryo using interspecific chimaeras between the rat and mouse. In: Cell Patterning (CIBA Foundation Symposium 29), pp. 183–200. Associated Scientific Publishers, Amsterdam.

Gardner, R. L., and Papaioannou, V. E. (1975). Differentiation in trophectoderm and inner cell mass. In: The Early Development of Mammals, pp. 107–132. (Balls, M., and Wild, A. E., eds.) Cambridge University Press, London.

Gardner, R. L., and Rossant, J. (1976). Determination during embryogenesis. In: Embryogenesis in Mammals (CIBA Foundation Symposium 40, New Series)), pp. 5–25. Elsevier, Amsterdam.

Gilula, N. B. (1974). Junctions between cells. In: Cell Communication (Cox, R. E., ed.). J. Wiley & Sons, Inc., New York.

Gilula, N. B., Reeves, O. R., and Steinbach, A. (1972). Metabolic coupling, ionic coupling and cell contact. Nature (London) 235, 262–265.

Gilula, N. B., Epstein, M. L., and Beers, W. H. (1978). Cell-to-cell communication and ovulation. A study of the cumulus-oocyte complex. J. Cell Biol. 78, 58–75.

Gondos, B., and Bhiraleus, P. (1970). Pronuclear relationship and association of maternal and paternal chromosomes in flushed rabbit ova. Z. Zellforsch. mikrosk. Anat. 111, 149–159.

Gordon, V. M., Fraser, L. R., and Dandekar, P. V. (1975). The effect of ruthenium red and Concanavalin A on the vitelline surface of fertilized and unfertilized rabbit ova. Anat. Rec. 181, 95–112.

Gould, K. G. (1975). Mammalian fertilization. In: Scanning Electron Microscopic Atlas of Mammalian Reproduction (Hafez, E. S. E., ed.). Springer-Verlag Berlin.

Graaf, R. de (1672). De mulierum organis generatoni inservientibus tractatus novus, demonstrans tam homines et animalia, catera omnia. quae vivipara dicuntur, haud minus quam ovipara, ab ovo. Originem ducereo. Leiden.

Graham, C. F. (1973). Nucleic acid metabolism during early mammalian development. In: Regulation of Mammalian Reproduction, pp. 286–301 (Segal, S., Crozier, R., Corfman, P. A., and Condliffe, P. G., eds.). Charles C. Thomas, Springfield, Ill.

Graham, C. F. (1974). The production of parthenogenetic mammalian embryos and their use in biological research. Biol. Rev. 49, 399–422.

Grant, P., Nilsson, B. O., and Bergstrom, S. (1977). The ultrastructure of preimplantation hamster blastocysts developed in vivo and in vitro. Fert. Steril. 28, 866–870.

Griepp, E. B., and Revel, J. P. (1977). Gap junctions in development. In: Intracellular Communication (De Mello, W. C., ed.). Plenum Press, New York and London.

Gulyas, B. J. (1971a). The rabbit zygote: Formation of annulate lamellae. J. Ultrastruct. Res. 35, 112–126.

Gulyas, B. J. (1971b). Nuclear extrusion in rabbit embryos. Z. Zellforsch. mikrosk. Anat. 120, 151–159.

Gulyas, B. J. (1972). The rabbit zygote. III. Formation of the blastomere nucleus. J. Cell Biol. 55, 533–541.

Gulyas, B. J. (1974a). Electron microscopic observations on advanced stages of spontaneous polyspermy in rabbit zygotes. Anat. Rec. 179, 285–296.

Gulyas, B. J. (1974b). Cortical granules in artificially activated (parthenogenetic) rabbit eggs. Am. J. Anat. 140, 577–582.

Gwatkin, R. B. L. (1976). Fertilization. In: The Cell Surface in Animal Embryogenesis and Development (Poste, G., and Nicolson, G. L., eds.). North Holland Publishing Co., Amsterdam and New York.

Gwatkin, R. B. L. (1977). Fertilization Mechanisms in Man and Mammals. Plenum Press, New York and London.

Gwatkin, R. B. L., Williams, D. T., Hartmann, J. F., and Kniazuk, M. (1973). The zona reaction of hamster and mouse eggs: Production in vitro by a trypsinlike protease from cortical granules. J. Reprod. Fert. 32, 259–265.

Gwatkin, R. B. L., Rasmusson, G. H., and Williams, W. T. (1976). Induction of the cortical reaction in hamster eggs by membrane-active agents. J. Reprod. Fert. 47, 299–303.

Hadek, R., and Swift, H. (1960). A crystalloid inclusion in the rabbit blastocyst. J. Biophys. Biochem. Cytol. 8, 836–841.

Hadek, R., and Swift, H. (1962). Nuclear extrusion and intracisternal inclusions in the rabbit blastocyst. J. Cell Biol. 13, 445–451.

Handyside, A. H., and Barton, S. C. (1977). Evaluation of the techniques and immunosurgery for the isolation of inner cell masses from mouse blastocysts. J. Embryol. Expl. Morph. 37, 217–226.

Hastings, R., and Enders, A. C. (1974). Uptake of exogenous proteins by the preimplantation rabbit. Anat. Rec. 179, 311–330.

Hastings, R., and Enders, A. C. (1975). Junctional complexes in the preimplantation rabbit embryo. Anat. Rec. 181, 17–34.

Hegner, R. W. (1914). Studies on germ cells. J. Morphol. 25, 375–510.

Herbert, M. C., and Graham, C. F. (1974). Cell determination and biochemical differentiation of the early embryo. In: Current Topics in Developmental Biology, Vol. 8. Academic Press, New York.

Hillman, N., and Tasca, R. J. (1969). Ultrastructural and autoradiographic studies of mouse cleavage stages. Am. J. Anat. 126, 151–174.

Hillman, N., Sherman, M. I., and Graham, C. F. (1972). The effect of spatial arrangement on cell determination during mouse development. J. Embryol. Exp. Morph. 28, 263–278.

Hoffman, L. H., Davies, J., and Long, V. D. (1975). Hormone induced crystals and intramitochondrial lamellae in uterine epithelium. In: Electron Microscopic Concepts of Secretion (Gess, M., ed.). John Wiley & Sons, New York.

Huebner, R. J., Kelloff, G. J., Sarma, P. S., Lane, W. T., Turner, A. C., Gilden, R. V., Oroszlan, S., Merer, H., Myers, D. B., and Peters, R. L. (1970). Group-specific antigen expression during embryogenesis of the genome of the C-type RNA tumor virus: Implications for ontogenesis and oncogenesis. Proc. Nat. Acad. Sci. (USA) 67, 366–376.

Illmensee, K., and Mahowald, A. P. (1974). Transplantation of posterior polar plasm in Drosophila. Induction of germ cells at the anterior pole of the eggs. Proc. Nat. Acad. Sci (USA) 71, 1016–1020.

Inoue, M., and Wolf, D. P. (1975). Comparative solubility properties of the zonae pellucidae of unfertilized and fertilized mouse ova. Biol. Reprod. 11, 558–565.

Jaenisch, R., and Berns, A. (1977). Tumor virus expression

during mammalian embryogenesis. In: Concepts in Mammalian Embryogenesis (Sherman, M. I., ed.). M. I. T. Press, Cambridge, Mass. and London.

Johnson, M. H. (1975). The macromolecular organization of membranes and its bearing on events leading up to fertilization. J. Reprod. Fert. *44*, 167–184.

Johnson, M. H., and Howe, C. W. S. (1975). The immunobiology of spermatozoa. In: The Biology of the Male Gamete (Duckett, J. G., and Racey, P. A., eds.). Biol. J. Linnean Soc. 7, suppl. 1, 205–214.

Johnson, M. H., Eager, D., Muggleton-Harris, A., and Grave, H. M. (1975). Mosaicism in organization of concanavalin A receptors on surface membrane of mouse egg. Nature (London) *257*, 321–322.

Kalter, S. S., Helmke, R. J., Heberling, R. L., Panigel, M., Fowler, A. K., Strickland, J. E., and Hellman, A. (1973). C-type particles in normal human placentas. J. Nat. Cancer Inst. *50*, 1081–1084.

Kalter, S. S., Helmke, R. J., Panigel, M., Heberling, R. L., Felsburg, P. J., and Axelrod, L. R. (1973). Observations of apparent C-type particles in baboon *(Papio cynocephalus)* placentas. Science *179*, 1332–1333.

Kalter, S. S., Panigel, M., Kraemer, D. C., Heberling, R. L., Helmke, R. J., Smith, G. C., and Hellman, A. (1974). C-type particles in baboon *(Papio cynocephalus)* preimplantation embryos. J. Natl. Cancer Inst. *52*, 1927–1928.

Karp, G., Manes, C., and Hahn, W. E. (1973). RNA synthesis in the preimplantation rabbit embryo: Radioautographic analysis. Develop. Biol. 31, 404–408.

Kaufman, M., and Sachs, L. (1976). Complete preimplantation development in culture of parthenogenetic mouse embryos. J. Embryol. Expl. Morph. *35*, 179–190.

Kaufman, M. E., Barton, S. C., and Surani, M. A. H. (1977). Normal postimplantation development of mouse parthenogenetic embryos to the forelimb bud stage. Nature (London) *265*, 53–55.

Kessel, R. G. (1969). Cytodifferentiation in the *Rana pipiens* oocyte. I. Association between mitochondria and nucleolus-like bodies in young oocytes. J. Ultrastruct. Res. *28*, 61–75.

Knowland, J., and Graham, C. F. (1972). RNA synthesis in the two-cell stage of mouse development. J. Embryol. Exp. Morph. *27*, 167–176.

Lewis, W. H., and Gregory, P. W. (1939). Cinematography of living rabbit eggs. Science *69*, 226–229.

Lin, T. P. (1969). Microsurgery of the inner cell mass of mouse blastocysts. Nature (London) *222*, 480–481.

Longo, F. J. (1973). Sperm aster in rabbit zygotes: Its structure and function. J. Cell Biol. *69*, 539–547.

Longo, F. J., and Anderson, E. (1969). Cytological events leading to the formation of the two-cell stage in the rabbit: Association of the maternally and paternally derived genomes. J. Ultrastruct. Res. *29*, 86–118.

Mahi, C. A., and Yanagimachi, R. (1975). Induction of nuclear decondensation of mammalian spermatozoa *in vitro*. J. Reprod. Fert. *44*, 293–296.

Mahowald, A. P. (1962). Fine structure of pole cells and polar granules in *Drosophila melanogaster*. J. Exp. Zool. *151*, 201–215.

Mahowald, A. P. (1968). Polar granules in *Drosophilia*. II. Ultrastructural changes during early embryogenesis. J. Exp. Zool. *167*, 237–262.

Mahowald, A. P. (1971). Origin and continuity of polar granules. In: Origin and Continuity of Cell Organelles (Reinert, J., and Ursprung, eds.). Springer-Verlag, Berlin.

Mahowald, A. P., and Hennen, S. (1971). Ultrastructure of the "germ plasm" in eggs and embryos of *Rana pipiens*. Develop. Biol. *24*, 37–53.

Manes, C. (1971). Nucleic acid synthesis in preimplantation rabbit embryos. II. Delayed synthesis of ribosomal RNA. J. Exp. Zool. *176*, 87–95.

Manes, C. (1973). The participation of the embryonic genome during early cleaving in the rabbit. Develop. Biol. *32*, 453–459.

Manes, C. (1975). Genetic and biochemical activities in preimplantation embryos. In: The Developmental Biology of Reproduction (Markert, C. L., and Papaconstantinou, J., eds.). Academic Press, New York and London.

Manes, C. (1977). Nucleic acid synthesis in preimplantation rabbit embryos. III. A "dark period" immediately following fertilization, and the early predominance of low molecular weight RNA synthesis. J. Exp. Zool. *201*, 247–258.

Manzanek, K. (1965). Submikroskopische Veränderungen während der Furehung eines Saugetieries. Arch. Biol. Liege 7, 49–85.

Mazia, D. (1937). The release of calcium in *Arbacia* eggs upon fertilization. J. Cell. Comp. Physiol. 10, 291–304.

McGaughey, R. W., and Van Blerkom, J. (1977). Patterns of polypeptide synthesis of porcine oocytes during maturation *in vitro*. Develop. Biol. *56*, 241–254.

McMahon, D. (1974). Chemical messengers in development: A hypothesis. Science *185*, 1012–1021.

McNutt, N. S., and Weinstein, R. S. (1973). Membrane ultrastructure at mammalian intercellular junctions. Prog. Biophys. Mol. Biol. *26*, 45–101.

McReynolds, H. D., and Hadek, R. (1972). A comparison of the fine structure of late mouse blastocysts developed *in vivo* and *in vitro*. J. Exp. Zool. *182*, 95–118.

Meyers, H. I., Young, W. C., and Dempsey, E. W. (1936). Graafian follicle development throughout the reproductive cycle in the guinea pig with especial reference to the changes during oestrous and sexual receptivity. Anat. Rec. *65*, 381–401.

Miller, L., and Gonzales, F. (1976). The relationship of ribosomal RNA synthesis to the formation of segregated nucleoli and nucleolus-like bodies. J. Cell Biol. *71*, 939–949.

Mills, R. M., and Brinster, R. L. (1967). Oxygen consumption of preimplantation mouse embryos. Expl. Cell Res. *47*, 337–344.

Morton, D. B. (1975). Acrosomal enzymes: Immunochemical localisation of acrosin and hyaluronidase in ram spermatozoa. J. Reprod. Fert. *45*, 375–378.

Morton, D. B. (1977). Immunoenzymatic studies on acrosin and hyaluronidase in ram spermatozoa. In: Immunobiology of the Gametes (Edidin, M., and Johnson, M. H., eds.). Cambridge University Press, Cambridge, England.

Moskalewski, S., Sawicki, W., Gabara, B., and Koprowski, H. (1971). Crystalloid formation in unfertilized mouse ova under influence of cytochalasin B. J. Exp. Zool. *180*, 1–12.

Motta, P., and Van Blerkom, J. (1974). Presence d'un materiel characteristique granulaire dans le cytoplasma de l'ovocytes et dans les premiers stades de la differenciation des cellules embryonaires. Bull. Assoc. Anat. Fr. *58*, 350–355.

Muggleton-Harris, A. L., and *Johnson, M. H.* (1976). The nature and distribution of serologically detectable alloantigens on the preimplantation mouse embryo. J. Embryol. Exp. Morph. *35*, 59–72.

Nako, K., Meyer, C. J., and *Noda, Y.* (1971). Progesterone-specific protein crystals in the endometrium: An electron microscopic study. Amer. J. Obstet. Gynecol. *111*, 1034–1038.

Nicolson, G. L., Yanagimachi, R., and *Yanagimachi, H.* (1975). Ultrastructural localization of lectin-binding sites on the zona pellucida and plasma membranes of mammalian eggs. J. Cell Biol. *66*, 263–274.

Nicosia, S. V., Wolf, D. P., and *Inoue, M.* (1977). Cortical granule distribution and cell surface characteristics in mouse eggs. Devel. Biol. *57*, 56–74.

Nicosia, S. V., Wolf, D. P., and *Mastroianni, L.* (1978) Surface topography of mouse eggs before and after insemination. Devel. Biol. (submitted).

Noda, Y. D., and *Yanagimachi, R.* (1976). Electron microscopic observations of guinea pig spermatozoa penetrating eggs *in vitro.* Development, Growth and Differentiation *18*, 15–23.

Odor, D. L. (1960). Electron microscopic studies on ovarian oocytes and unfertilized tubal ova in the rat. J. Biophys. Biochem. Cytol. *7*, 567.

Odor, L. D., and *Renninger, D. F.* (1960). Polar body formation in the rat oocyte as observed with the electron microscope. Anat. Rec. *137*, 13–23.

Oikawa, T., Yanagimachi, R., and *Nicolson, G. L.* (1973). Wheat germ agglutinin blocks mammalian fertilization. Nature (London) *241*, 256–259.

Oprescu, St., and *Thibault, C.* (1965). Duplication de l'ADN dans les oeufs de lapine après la fécondation. Ann. Biol. Anim. Biochim. Biophys. *5*, 151–156.

Overstreet, J. W., and *Bedford, J. M.* (1974). Comparison of the permeability of the egg vestments in follicular oocytes, unfertilized and fertilized ova of the rabbit. Devel. Biol. *41*, 185–192.

Panigel, M., Kraemer, D. C., Kalter, S. S., Smith, F. C. and *Heberling, R. L.* (1975). Ultrastructure of cleavage stage and preimplantation embryos of the baboon. Anat. Embryol. *147*, 45–62.

Pikó, L. (1969). Gamete structure and sperm entry in mammals. In: Fertilization, Vol. 2, pp. 325–403 (Metz, C., and Monroy, A., eds.). Academic Press, New York.

Pikó, L. (1977). Immunocytochemical detection of a murine leukemia virus related nuclear antigen in mouse oocytes and early embryos. Cell *12*, 697–707.

Pikó, L., and *Tyler, A.* (1964). Fine structural studies of sperm penetration in the rat. Proceedings of the 5th International Congress Animal Reproduction, Trento, vol. 2, pp. 372–377.

Sarkar, N. H., Moore, D. H., and *Nowinski, R. C.* (1972). Symmetry of the nucleocapsid of the oncornaviruses. In: RNA Viruses and Host Genome in Oncogenesis, pp. 71–79 (Emmelot, P., and Bentvelzen, P., eds.) North Holland Publishing Co., Amsterdam and New York.

Schidlovsky, G., and *Ahmed, M.* (1973). C-type virus particles in placentas and fetal tissues of Rhesus monkeys. J. Nat. Cancer Inst. *51*, 225–233.

Schlafke, S., and *Enders, A. C.* (1967). Cytological changes during cleavage and blastocyst formation in the rat. J. Anat. *102*, 13–32.

Schuchner, E. B. (1970). Ultrastructural changes of the nucleoli during early development of fertilized rat eggs. Biol. Reprod. *3*, 265–274.

Schultz, G. A. (1973). Characterization of polyribosomes containing newly synthesized messenger RNA in preimplantation rabbit embryos. Expl. Cell Res. *82*, 168–174.

Schultz, G. A. (1975). Polyadenylic acid containing RNA in unfertilized and fertilized eggs of the rabbit. Devel. Biol. *44*, 270–277.

Schultz, G. A., and *Church, R. B.* (1975). Transcriptional patterns in early mammalian development. In: Biochemistry of Animal Development, Vol. III (Weber, R., ed.). Academic Press, New York.

Schultz, G. A., and *Tucker, E. B.* (1977). Protein synthesis and gene expression in preimplantation rabbit embryos. In: Development in Mammals, Vol. 1 (Johnson, M. H., ed.). North Holland Publishing Co., Amsterdam and New York.

Smith, D., and *Williams, M. A.* (1975). Germinal plasm and determination of the primordial germ cells. In: The Developmental Biology of Reproduction (Markert, C. L., and Papaconstantinou, J., eds.). Academic Press, New York.

Snow, M. H. L. (1973). The differential effect of ^{3}H-thymidine upon two populations of cells in preimplantation mouse embryos. In: The Cell Cycle in Development and Differentiation, pp. 311–324 (Balls, M., and Billett, F. S., eds.). Cambridge University Press, Cambridge, England.

Snow, M. H. L., Aitken, J., and *Ansell, J. D.* (1976). Role of the inner cell mass in controlling implantation in the mouse. J. Reprod. Fert. *29*, 123–126.

Solter, D. (1977). Organization and the antigenic properties of the egg membrane. In: Immunobiology of the Gametes (Edidin, M., and Johnson, M. H., eds.). Cambridge University Press, Cambridge, England.

Solter, D., and *Knowles, B. S.* (1975). Immunosurgery of mouse blastocyst. Proc. Nat. Acad. Sci. (USA) *72*, 5099–5102.

Sotelo, J. R., and *Porter, K. R.* (1959). An electron microscopic study of the rat ovum. J. Biophys. Biochem. Cytol. *5*, 327–341.

Soupart, P., and *Strong, P. A.* (1974). Ultrastructural observations on human oocytes fertilized *in vitro.* Fertil. and Steril. *25*, 11–44.

Soupart, P., and *Strong, P. A.* (1975). Ultrastructural observations on polyspermic penetration of zona pellucida-free human oocytes inseminated *in vitro.* Fert. and Steril. *26*, 523–537.

Stefanini, M., Oura, C., and *Zamboni, L.* (1969). Ultrastructure of fertilization in the mouse. 2. Penetration of sperm into the ovum. J. Submicr. Cytol. *1*, 1–23.

Steinhardt, R. A., Epel, D., Carroll, D. J., and *Yanagimachi, R.* (1974). Is calcium ionophore a universal activator for unfertilized eggs? Nature (London) *252*, 41–43.

Stern, S., Biggers, J., and *Anderson, E.* (1971). Mitochondria and early development of the mouse. J. Expl. Zool. *176*, 179–192.

Stevens, L. C., Varnum, D. S., and *Eicher, E. M.* (1977). Viable chimeras produced from normal and parthenogenetic mouse embryos. Nature (London) *269*, 515–517.

Strand, M., August, T., and Jaenisch, R. (1977). Oncornavirus gene expression during embryonal development of the mouse. Virology 76, 886–890.

Sugawara, S., Takeuchi, S., and Hafez, E. S. E. (1975). Sperm penetration. In: Scanning Electron Microscopic Atlas of Mammalian Reproduction (Hafez, E. S. E., ed.). Springer-Verlag, Berlin.

Surani, M. A. H., and Barton, S. C. (1977). Trophoblastic vesicles of preimplantation blastocysts can enter into quiescence in the absence of inner cell mass. J. Embryol. Exp. Morph. 39, 273–277.

Szollosi, D. (1965a). Development of "yolky substance" in some rodent eggs. Anat. Rec. 151, 424 (abs.).

Szollosi, D. (1965b). Extrusion of nucleoli from pronuclei of the rat. J. Cell Biol. 25, 545–562.

Szollosi, D. (1966). Time and duration of DNA synthesis in rabbit eggs after sperm penetration. Anat. Rec. 154, 202–212.

Szollosi, D. (1967). Development of cortical granules and the cortical reaction in rat and hamster eggs. Anat. Rec. 159, 431–446.

Szollosi, D. (1971). Nucleoli and ribonucleoprotein particles in the preimplantation conceptus of the rat and mouse. In: The Biology of the Blastocyst (Blandau, R. J., ed.). University of Chicago Press, Chicago and London.

Szollosi, D. G., and Ris, H. (1961). Observations on sperm penetration in the rat. J. Biophys. Biochem. Cytol. 10, 275–283.

Tarkowski, A. K., and Wroblewska, J. (1967). Development of blastomeres of mouse eggs isolated at the 4- and 8-cell stage. J. Embryol. Exp. Morph. 18, 155–180.

Tarkowski, A. K., Witkowska, A., and Nowicka, J. (1970). Experimental parthenogenesis in the mouse. Nature (London) 266, 162–165.

Thibault, C. (1973). In vitro maturation and fertilization of rabbit and cattle oocytes. In: The Regulation of Mammalian Reproduction, pp. 231–240 (Segal, S. J., Crozier, R., Corfman, P. A., and Condliffe, P. G., eds.). Charles C. Thomas, Springfield, Ill.

Thibault, C. (1977). Are follicular maturation and oocyte maturation independent processes? J. Reprod. Fert. 51, 1–15.

Thibault, C., and Gerard, M. (1973). Cytoplasmic and nuclear maturation of rabbit oocytes in vitro. Ann. Biol. Anim. Biochim. Biophys. 13 (suppl.) 145–156.

Thompson, R. S., Moore-Smith, D., and Zamboni, L. (1974). Fertilization of mouse ova in vitro: An electron microscopic study. Fertil. Steril. 25, 222–249.

Todaro, G., and Huebner, R. (1972). The viral oncogene hypothesis: New evidence. Proc. Nat. Acad. Sci. (USA) 69, 1009–1015.

Tyndale-Biscoe, C. H. (1965). Fine structure of the rabbit blastocyst. Aust. Mammal Soc. Bull. 2, 38–39.

Uehara, T., and Yanagimachi, R. (1976). Microsurgical injection of spermatozoa into hamster eggs with subsequent transformation of sperm nuclei into male pronuclei. Biology of Reproduction 15, 467–470.

Usui, N., and Yanagimachi, R. (1976). Behavior of hamster sperm nuclei incorporated into eggs at various stages of maturation, fertilization and early development. J. Ultrastruct. Res. 57, 276–288.

Van Beneden, E. (1880). Recherches sur l'embryologie des mammifères: La formation des feuillets chez le lapin.

Arch. Biol. (Liège) 1, 137.

Van Blerkom, J. (1977). Molecular approaches to the study of oocyte maturation and embryonic development. In: Immunobiology of the Gametes (Edidin, M., and Johnson, M. H., eds.). Cambridge University Press, Cambridge, England.

Van Blerkom, J., Manes, C., and Daniel, J. C., Jr. (1973). Development of preimplantation rabbit embryos in vivo and in vitro. I. An ultrastructural comparison. Devel. Biol. 35, 262–282.

Van Blerkom, J., and Manes, C. (1974). Development of preimplantation rabbit embryos in vivo and in vitro. II. A comparison of qualitative aspects of protein synthesis. Devel. Biol. 40, 40–51.

Van Blerkom, J., and Brockway, G. (1975). Qualitative patterns of protein synthesis in the preimplantation mouse embryo. I. Normal pregnancy. Devel. Biol. 44, 148–157.

Van Blerkom, J., Barton, S. C., and Johnson, M. H. (1976). Molecular differentiation in the preimplantation mouse embryo. Nature (London) 259, 319–321.

Van Blerkom, J., and Runner, M. N. (1976). The fine structural development of preimplantation mouse parthenotes. J. Exp. Zool. 196, 113–123.

Van Blerkom, J., and Manes, C. (1977). The molecular biology of the preimplantation embryo. In: Concepts in Mammalian Embryogenesis (Sherman, M. H., ed.), pp. 37–94. M. I. T. Press, Cambridge, Mass.

Van Blerkom, J., and McGaughey, R. W. (1978a). Molecular differentiation of the rabbit ovum. I. During the in vivo and in vitro maturation of the oocyte. Devel. Biol. 63, 139–150.

Van Blerkom, J., and McGaughey, R. W. (1978b). Molecular differentiation of the rabbit ovum. II. During the preimplantation development of in vivo and in vitro matured oocytes. Devel. Biol. 63, 151–164.

Vernon, M. L., McMahon, J. M., and Hackett, J. J. (1974). Additional evidence of type-C particles in human placentas. J. Nat. Cancer Inst. 52, 987–989.

von Baer, K. E., (1827). De ovi mammalium et hominis genesi. Leipzig.

von Baer, K. E. (1828). Über Entwickelungsgeschichte der Thiere. Beobachtung und Reflexion. Königsberg.

Wales, R. G. (1970). Effects of ions on the development of the preimplantation mouse embryo in vitro. Aust. J. Biol. Sci. 23, 421–429.

Weakley, B. S. (1966). Electron microscopy of the oocyte and granulosa cells in the developing ovarian follicles of the golden hamster (Mesocricetus auratus). J. Anat. 100, 503–534.

Weakley, B. S. (1967). Investigations into the structure and fixation properties of cytoplasmic lamellae in the hamster oocyte. Z. Zellforsch. mikrosk. Anat. 81, 91–99.

Weakley, B. S. (1968). Comparison of cytoplasmic lamellae and membranous elements in the oocytes of five mammalian species. Z. Zellforsch. mikrosk. Anat. 85, 109–123.

Wiley, L. M., and Calarco, P. G. (1975). The effects of anti-embryo sera and their localization on the cell surface during mouse preimplantation development. Devel. Biol. 47, 407–418.

Williams, M. A., and Smith, L. D. (1971). Ultrastructure of the 'germinal plasm' during maturation and early cleavage in Rana pipiens. Devel. Biol. 25, 568–580.

Wilson, E. B. (1925). The Cell in Development and Heredity. Macmillan Co., New York.

Wischnitzer, S. (1967). Intramitochondrial transformations during oocyte maturation in the mouse. J. Morph. *121*, 29–46.

Witkowska, A. (1973a). Parthenogenetic development of mouse embryos *in vivo*. 1. Preimplantation development. J. Embryo Exptl. Morph. *30*, 519–545.

Witkowska, A. (1973b). Parthenogenetic development of mouse embryos *in vivo*. 2. Postimplantation development. J. Embryo Exptl. Morph. *30*, 547–560.

Wolf, D. P. (1974). The cortical response in *Xenopus laevis*. Proc. Nat. Acad. Sci. (USA) *71*, 2067–2071.

Yanagimachi, R. (1972). Fertilization of guinea pig eggs *in vitro*. Anat. Rec. *174*, 9–20.

Yanagimachi, R. (1977). Speceficity of sperm-egg interaction. In: The Immunobiology of the Gametes (Edidin, M., and Johnson, M. H., eds.). Cambridge University Press, Cambridge, England.

Yanagimachi, R., and Chang, M. C. (1961). Fertilizable life of golden hamster ova and their morphological changes at the time of losing fertilizability. J. Exptl. Zool. *148*, 185–204.

Yanagimachi, R., and Noda, Y. D. (1970a). Ultrastructural changes in the hamster sperm head during fertilization. J. Ultrastruct. Res. *31*, 465–485.

Yanagimachi, R., and Noda, Y. D. (1970b). Fine structure of the hamster sperm head. Amer. J. Anat. *128*, 367–388.

Yanagimachi, R., and Noda, Y. D. (1970c). Electron microscopic studies of sperm incorporation into the golden hamster egg. Amer. J. Anat. *128*, 429–462.

Yanagimachi, R., and Noda, Y. D. (1970d). Physiological changes in the postnuclear cap region of mammalian spermatozoa: A necessary preliminary to the membrane fusion between sperm and egg cells. J. Ultrastruct. Res. *31*, 486–493.

Yanagimachi, R., Yanagimachi, H., and Rogers, B. J. (1976). The use of zona-free animal ova as a test-system for the assessment of fertilizing capacity of human spermatozoa. Biol. Reprod. *15*, 471–476.

Zamboni, L. (1970). Ultrastucture of mammalian oocytes and ova. Biol. Reprod., suppl. *2*, 44–63.

Zamboni, L. (1971). Fine Morphology of Mammalian Fertilization. Harper and Row, New York.

Zamboni, L. (1972). Fertilization in the mouse. In: Biology of Mammalian Fertilization and Implantation, pp. 213–262 (Moghissi, K. S., and Hafez, E. S. E., eds.). Charles C. Thomas, Springfield, Ill.

Zamboni, L., and Mastroianni, L. (1966). Electron microscopic studies on rabbit ova. I. The follicular oocyte. J. Ultrastruct. Res. *14*, 95–117.

Plate 56. Mouse Oocyte with the First Polar Body.

A   Scanning electron micrograph of a mouse oocyte undergoing meiotic maturation. Notice in particular the contrast in surface appearance between the oocyte and the first polar body (Pb_1). While the oocyte surface is covered by microvilli, the surface of the polar body is quite smooth owing to the absence of microvilli or any other surface protrusions. The zona pellucida was removed enzymatically prior to fixation. (x 1,570).

B   A transmission electron micrograph of a mouse oocyte at a similar stage of meiotic maturation as shown in A reveals the contrasting surface characteristics of the oocyte and first polar body, especially in regard to the differential distribution of microvilli (Mv). Of particular interest in this micrograph is the appearance of a cytoplasmic bridge or "midbody" (Mb) between the oocyte (Oo) and the first polar body (Pb_1). (x 7,800).

C   Higher magnification view of the "midbody" presented in B. (x 11,900).

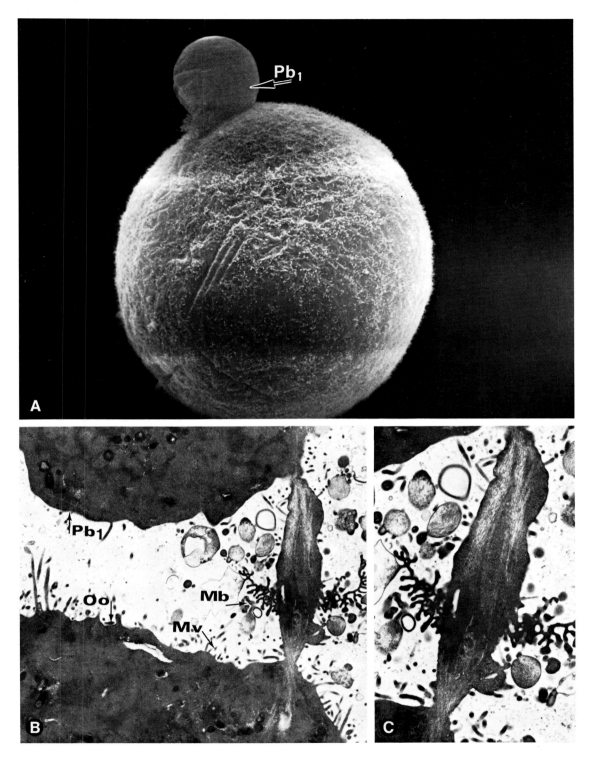

Plate 56. Mouse Oocyte with the First Polar Body.

Plate 57. Appearance of a Mouse Oocyte during the Formation of the Second Polar Body.

A   As is shown in this scanning electron micrograph, the surface of an oocyte that overlies the second metaphase spindle is devoid of microvilli (arrow). In fact, a sharp interface is present between this region and the remainder of the oocyte, which displays a dense population of microvilli. (x 27,000).

B   When observed by transmission electron microscopy, an oocyte in a similar stage of meiotic maturation as that shown in A contains numerous cortical Golgi complexes (G) which have been implicated in the formation of cortical granules (Cg). M = mitochondria. (x 21,600).

C   Deeper within the region that contains the second metaphase spindle are the chromosomes (Ch) and extensive accumulations of small vesicles or saccules (arrows) which eventually coalesce to reform the nuclear envelope of the female pronucleus. (x 28,000).

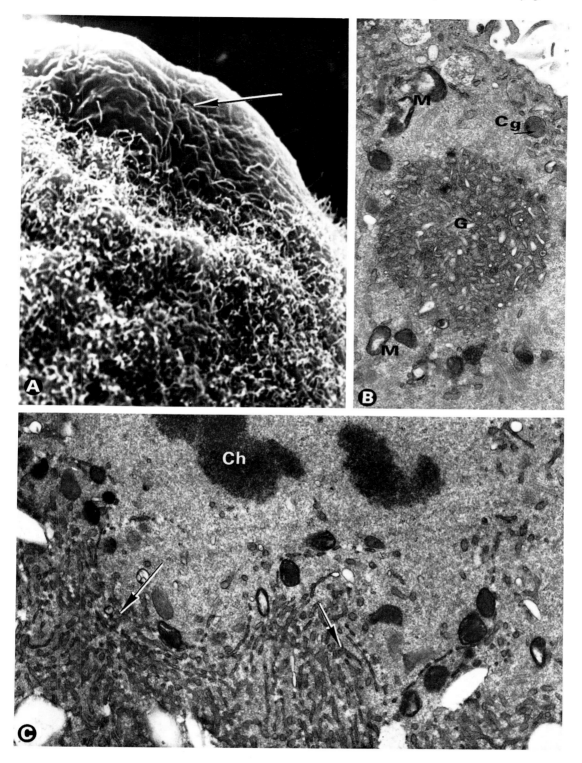

Plate 57. Appearance of a Mouse Oocyte during the Formation of the Second Polar Body.

Plate 58. The Abstriction of the Second Polar Body in the Mouse Oocyte.

A   Relatively low-magnification transmission electron microscopic view of the portion of the cytoplasm containing the second metaphase spindle in a mouse oocyte undergoing the final stages of meiotic maturation. In evidence in this region are numerous chromosomes (Ch) and the presence of small vesicles (v) which underlie the plasma membrane. Notice in particular the absence of microvilli (arrow *) and the smooth appearance of the plasma membrane. In addition, this particular cortical region is relatively devoid of cytoplasmic organelles. (x 15,400).

B   Scanning electron micrograph of a fertilized mouse egg in the process of extruding the second polar body (arrow). Note the smooth membrane of the polar body and the sharp line of division at the interface of the smooth and microvillous membranes. The tail section of the presumed fertilizing spermatozoon (Sp) is evident in this egg. (x 8,000).

C   Higher-magnification view of the surface of the extruding second polar body. Although the membrane is quite folded, it is clearly devoid of microvilli and other surface protrusions at this time. (x 20,000).

194

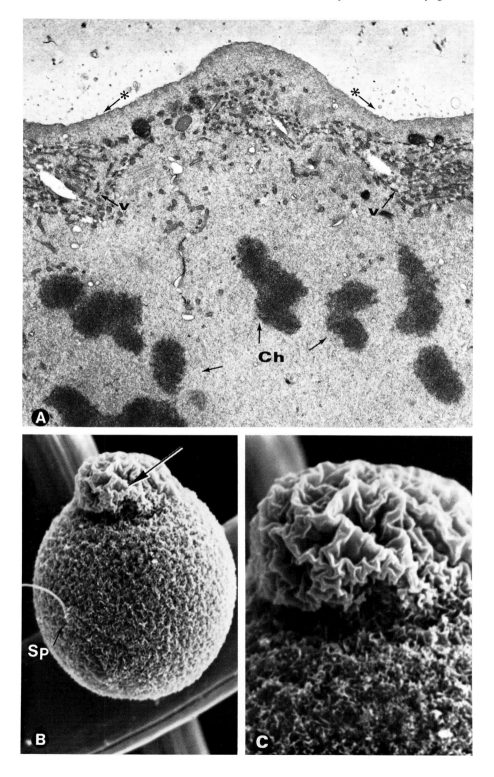

Plate 58. The Abstriction of the Second Polar Body in the Mouse Oocyte.

Plate 59. The Fertilized Rabbit Egg.

A   The apparent block to polyspermy in the rabbit is at the level of the vitelline membrane. After fertilization, numerous accessory spermatozoa (Sp) are observed in the perivitelline space (PVS) and in contact with the plasma membrane of the egg. Zp = zona pellucida; M = mitochondria. (x 5,200).

B   When examined at higher magnifications, accessory spermatozoa are frequently in association with the vitelline membrane (Vm), and the anterior portions of the head of spermatozoa are associated with small cytoplasmic protrusions of the egg (arrow). The spermatozoon indicated with an asterisk still retains a portion of its acrosomal cap. (x 49,000).

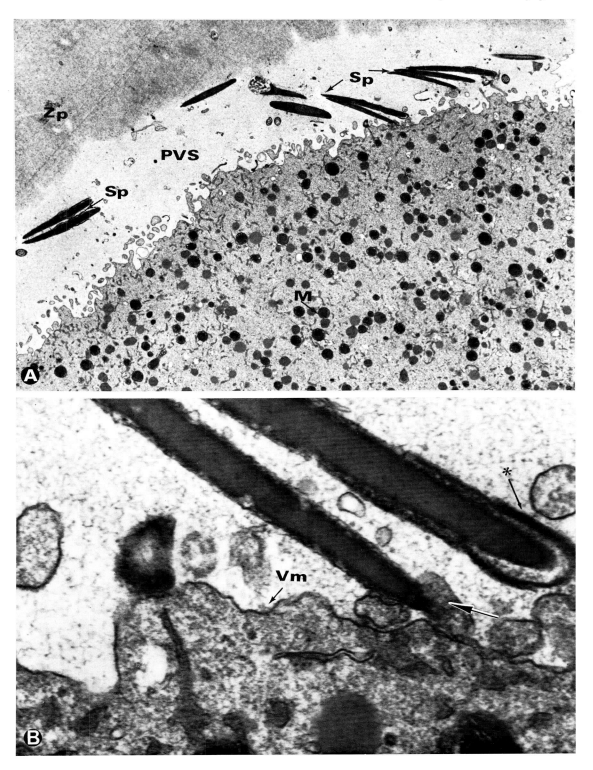

Plate 59. The Fertilized Rabbit Egg.

Plate 60. Cytoplasmic Features of the Early Mouse Embryo.

A   Two of the most characteristic features of the cytoplasm of a two-cell mouse embryo are virus-like particles and the presence of a dense population of lattice-like fibrils (Fb). The virus-like particles (Vp) are generally encountered within cisternae of the endoplasmic reticulum (intracisternal) and morphologically resemble A-type RNA viruses. N = nucleus. (x 25,500).

B   An additional feature of the cytoplasm of cleaving mouse embryos is the formation of numerous crystalloid bodies (Cr). These elements are frequently observed in association with cisternae of the endoplasmic reticulum (ER), suggesting to some investigators that the endoplasmic reticulum is involved in crystalloid formation. Fb = cytoplasmic fibrils. (x 24,000).

C   Frequently, crystalloids (Cr) and accumulations of virus-like particles (Vp) are observed in close association with the nucleus (N). (x 48,500).

D   The association of virus-like particles (Vp) and the nucleus (N) is most evident in this transmission electron micrograph. Of particular interest is the apparent continuity between the outer nuclear membrane (ONM) and the membrane that encloses the virus-like particles. This frequent observation has prompted some investigators to suggest a nuclear role in the formation and extrusion into the cytoplasm of these particles. (x 55,000).

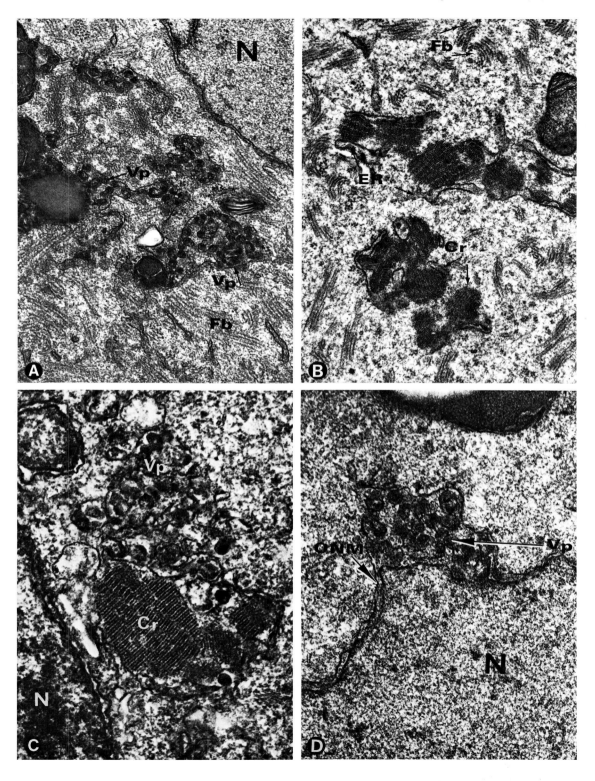

Plate 60. Cytoplasmic Features of the Early Mouse Embryo.

Plate 61. Virus-like Particles in Preimplantation Rabbit Embryos.

A   Virus-like particles (Vp) *morphologically* similar to A-type RNA viruses are occasionally observed in the cytoplasm of blastocyst-stage rabbit embryos. In contrast to the mouse, these particles are usually not membrane enclosed (intercisternal) but are typically associated with elements of the rough-surfaced endoplasmic reticulum (RER) and with masses of electron-dense material possessing a granular texture (arrow). (x 48,000).

B   Higher-magnification view of the virus-like particles shown in A. (x 110,000).

C, D An additional structure that is occasionally encountered in blastocyst stage rabbit embryos has an appearance that closely resembles C-type RNA viruses. These virus-like structures (Vp) in rabbit embryos are encountered either in the process of (apparently) being liberated from the cytoplasm (by budding from the plasma membrane) (C), or free within the intercellular spaces between adjacent cells (D). Whether these particles are actually RNA viruses is unknown at present. (C, x 97,000; D, x 120,000).

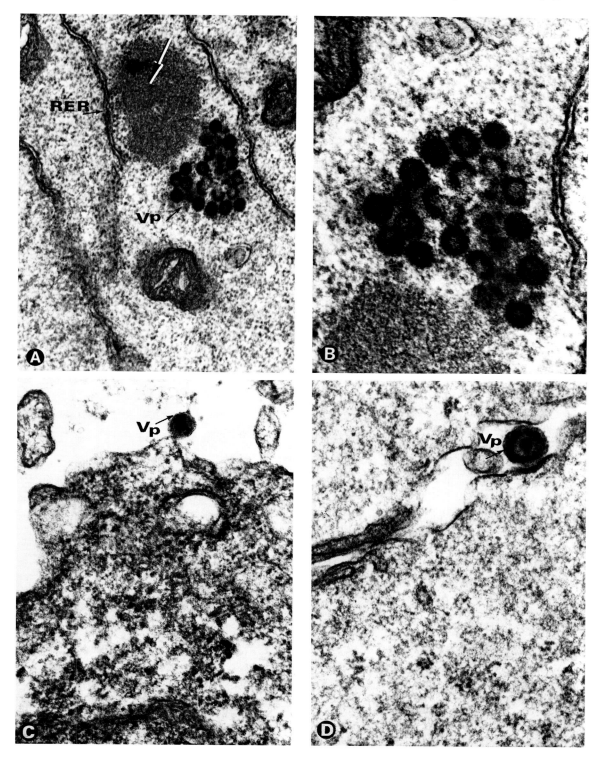

Plate 61. Virus-like Particles in Preimplantation Rabbit Embryos.

Plate 62. The Nuage.

A   In some insects and amphibians, cytoplasmic structures resembling nucleoli or appearing as electron-dense material located between mitochondria have been suggested to be determinants of germ cells. In the mammal, material of similar morphology, termed the nuage, has been observed by transmission electron microscopy. In this figure, nuage (Nu) is present between mitochondria (M) in a growing rabbit oocyte. (x 41,400).

B   In many mammalian oocytes, the nuage (Nu) is observed in a juxtanuclear position (N) and has an appearance similar to that of a nucleolus. (rabbit oocyte; x 38,600).

C   Cytoplasmic nuage (Nu) resembling a small nucleolus is shown in this transmission electron micrograph of a rabbit oocyte. M = mitochondria. (x 13,200).

D   The nuage (Nu) is also observed in some of the cells of preimplantation embryos, as is shown in this transmission micrograph taken from a 3.5-day-old rabbit embryo. It is unknown at present whether the nuage observed in mammalian oocytes or embryos has any role in the determination of the germ cells. N = nucleus. (x 23,500).

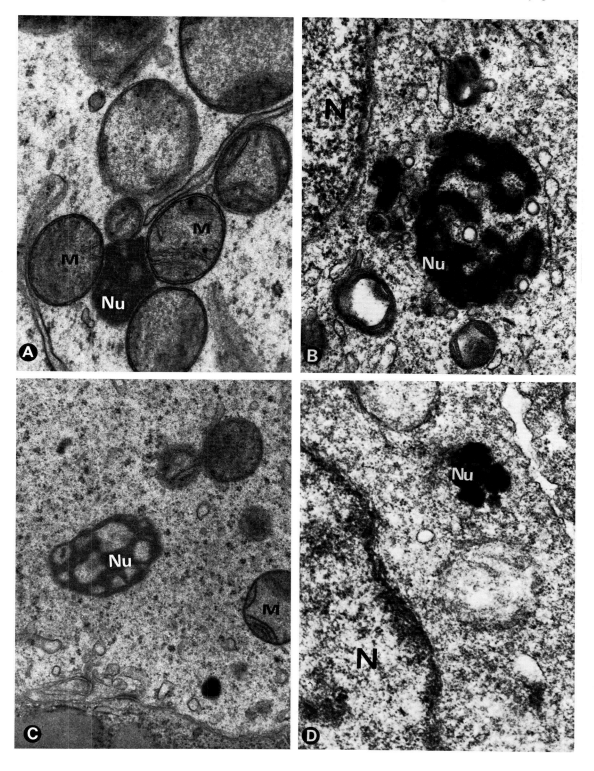

Plate 62. The Nuage.

Plate 63. Crystalloid Inclusions in Rabbit Trophoblast.

A   The most prominent cytoplasmic components of the rabbit trophoblast, visible during the expanded blastocyst stage, are numerous crystalline inclusions. Crystalline inclusions appear in the phase microscope as dense rods (Cr) and occupy the majority of the visible cytoplasm. N = nucleus. (x 1,100).

B   Transmission electron micrograph of a trophoblast cell in a 4.5-day-old rabbit embryo. Prominent in the cytoplasm are mitochondria (M) with matrices of low electron density and stacked, lamellar cristae. Portions of crystalline inclusions (Cr) are visible in the cytoplasm, which is also densely packed with polyribosomes (P). N = nucleus. (x 16,000).

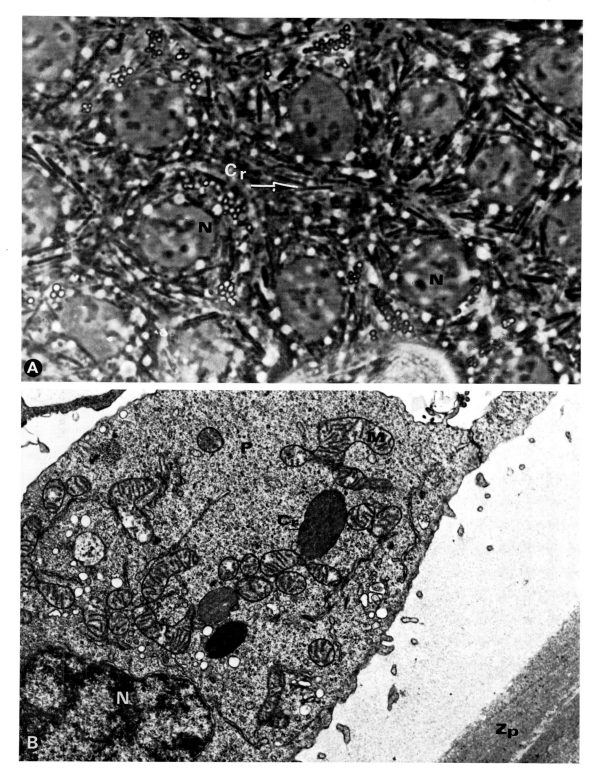

Plate 63. Crystalloid Inclusions in Rabbit Trophoblast.

Plate 64. Crystalline Inclusions in Rabbit Blastocysts.

A   A transmission electron microscopic comparison of the inner cell mass (ICM) and trophoblast (Tr) of a 5-day-old rabbit embryo reveals several special features of these cell types. First, crystalline inclusions (Cr) are only encountered in the trophoblast. Second, mitochondria in the inner cell mass cells are elongated and contain a matrix of moderate electron density and cristae that have a vacuolated appearance. Mitochondria in trophoblast cells are somewhat more polymorphic than in the inner cell mass and contain matrices of low to moderate electron density and stacked, lamellar cristae. N = nucleus. (x 9,800).

B   Higher-magnification transmission electron micrograph of the crystalline inclusion shown in A. Notice that the crystalline inclusion is membrane bound, and a regularly repeating periodicity is evident within the internal organization of the crystal. (x 80,000).

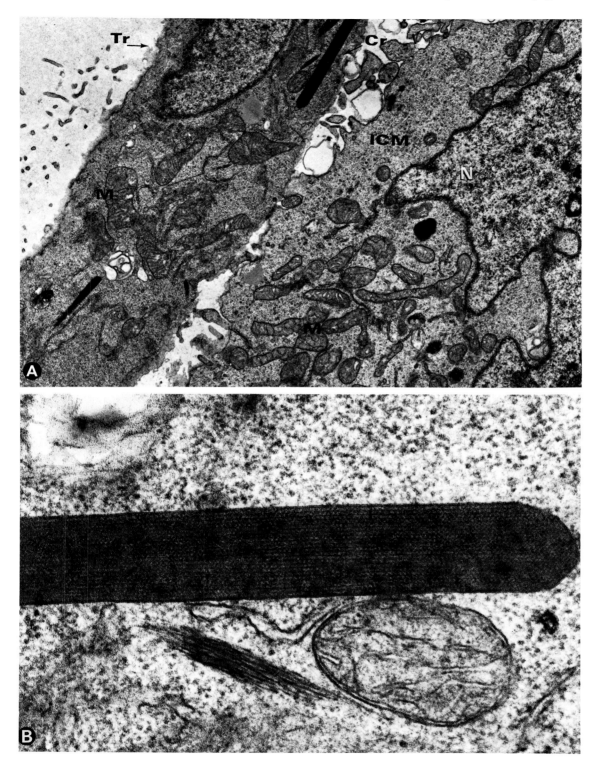

Plate 64. Crystalline Inclusions in Rabbit Blastocysts.

Plate 65. Nucleolar Maturation.

A   Nucleoli (nu) in newly fertilized rabbit eggs are electron-dense, spherical structures composed of a fibril-
     lar material. M = mitochondria. (x 11,300).

B   Nucleoli (nu) in the one-cell mouse embryo are morphologically quite similar to their counterparts in the
     rabbit. As in the rabbit, the mouse nucleus contains several spherical, electron-dense nucleoli of various
     sizes. (x 9,300).

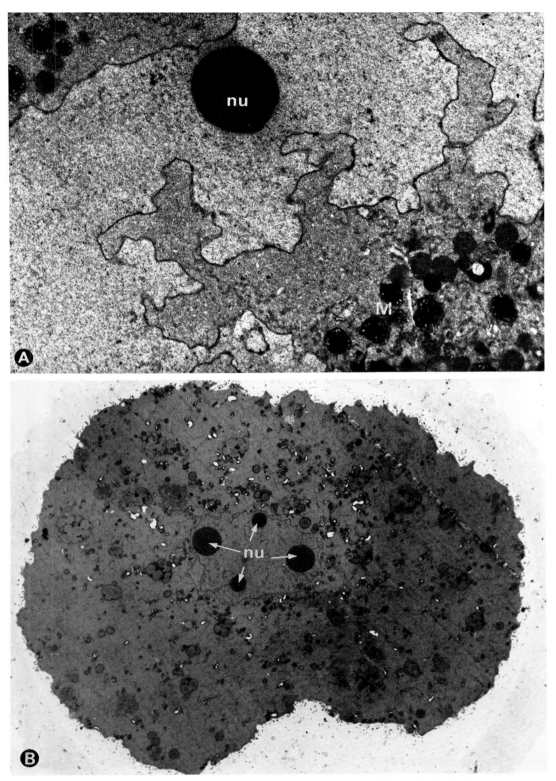

Plate 65. Nucleolar Maturation.

Plate 66. Nucleolar Maturation.

A   With continued preimplantation development of the mouse and rabbit embryo, major morphodynamic changes occur within the body of the nucleoli (nu). In this transmission electron micrograph, small holes (arrow) or "vacuoles" are evident within the nucleoli of a 1-day-old rabbit embryo. This is the first observable indication of nucleolar differentiation. (x 17,600).

B   By day 2, a granular element is present in rabbit nucleoli (g) and the more fibrillar (f) regions have begun to reticulate. (x 16,500).

C   In the early four-cell mouse embryo, nucleolar (nu) differentiation is initiated from the periphery of the structure, with a small granular component present in some of the reticulated zones. (x 19,200).

D   By the late four-cell stage, nucleoli (nu) in mouse embryos have completed a more advanced stage of reticulation, although the core of the structure remains unaltered in morphology. (x 10,000).

210

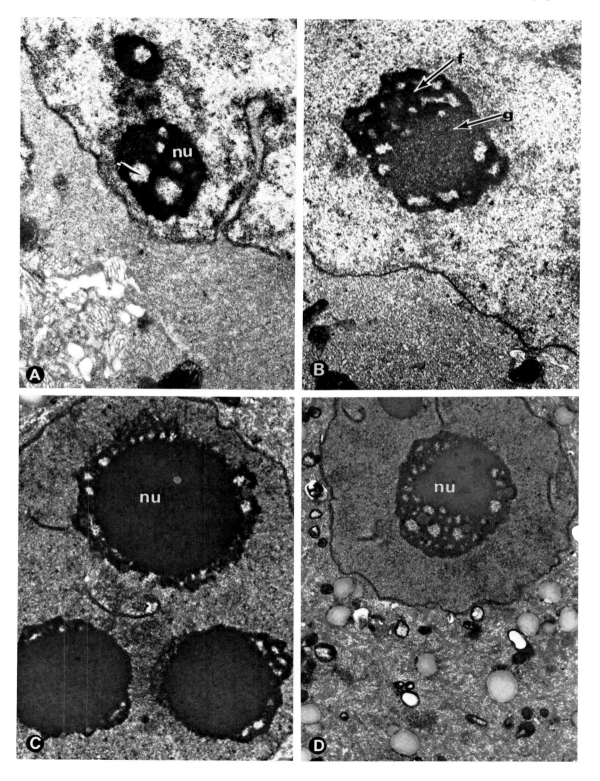

Plate 66. Nucleolar Maturation.

Plate 67. Appearance of Morphologically Mature Nucleoli in Preimplantation Rabbit and Mouse Embryos.

A   The characteristic appearance of a nucleolus in an early cavitating mouse blastocyst is presented in this transmission electron micrograph. Notice that the entire structure of the nucleolus (nu) is reticulated, especially the central portion. N = nucleus. (x 8,600).

B   Fully mature (differentiated) nucleoli (nu) in blastocyst-stage rabbit embryos are extensively reticulated and hypertropic structures in which granular and fibrillar elements are interspersed along anastomosing networks of nucleolonemas. N = nucleus. (x 8,800).

C   Fully differentiated nucleoli (nu) in "hatched" mouse blastocysts have the same general morphological appearance as is observed in the nucleoli of rabbit blastocysts. Tight junctions and other intercellular contacts between trophoblast cells (Tr) are indicated by an arrow with an asterisk. ICM = inner cell mass cells; BC = blastocyst cavity; N = nucleus. (x 2,600).

212

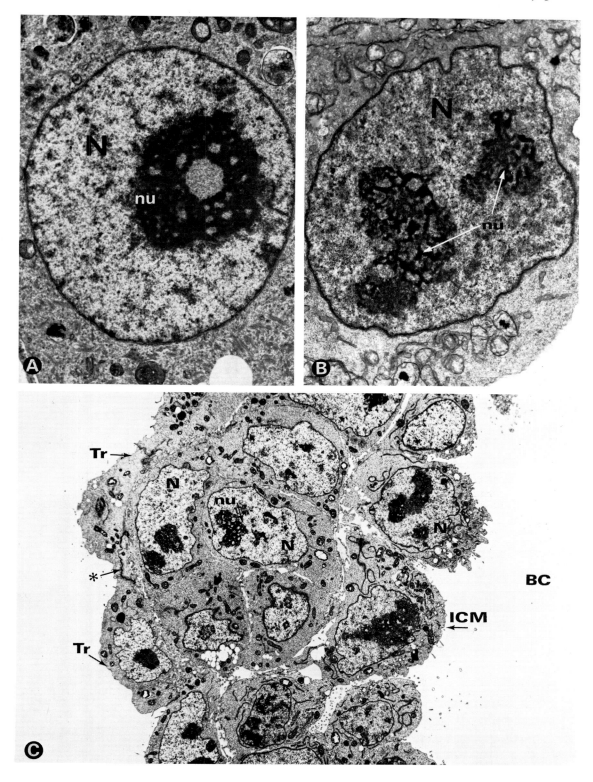

Plate 67. Appearance of Morphologically Mature Nucleoli in Preimplantation Rabbit and Mouse Embryos.

Plate 68. Mitochondrial Differentiation during Preimplantation Embryogenesis.

A   Mitochondria (M) in mature rabbit oocytes and newly fertilized eggs are comparatively small, spherical structures containing peripheral cristae that only partially penetrate matrices of high electron density. Also apparent in this particular transmission electron micrograph are flocculent vesicles (FV) and scattered clusters of ribosomes (P). (x 24,500).

B   By approximately day 2 of development, mitochondria (M) in cleaving rabbit embryos are somewhat more elongated than previously (M), but contain cristae that still only partially penetrate matrices of high electron density. (x 20,000).

C   Mitochondria in preimplantation mouse embryos are of the vacuolated type; i.e., they contain distended cristae that surround electron-translucent spaces. The matrices of mitochondria (M) in blastocyst-stage embryos are of a moderate electron density. In this transmission electron micrograph, two morphological forms of mitochondria are apparent: mitochondria in this particular trophoblast cell (Tr) contain stacked, lamellar cristae, while, by contrast, mitochondria in the inner cell mass cell (ICM) are of the more characteristic vacuolated type. (x 27,000).

D   This transmission electron micrograph of an inner cell mass cell of a hatched mouse blastocyst demonstrates the appearance of vacuolated mitochondria (M) and of distended cisternae of the rough-surfaced endoplasmic reticulum (RER). Notice that a material of moderate electron density is present within the cisternae. (x 25,000).

E   In this transmission electron micrograph of an inner cell mass cell of a 5-day-old rabbit blastocyst, an apparent continuity between the rough-surfaced endoplasmic reticulum (RER) and the outer nuclear membrane (ONM) is evident (arrow*). Since it is generally held that this organelle is derived from the nuclear membrane, this micrograph likely depicts the genesis of the rough-surfaced endoplasmic reticulum. The characteristic morphology displayed by mitochondria (M) in rabbit inner cell mass is also shown in this figure. (x 51,000).

214

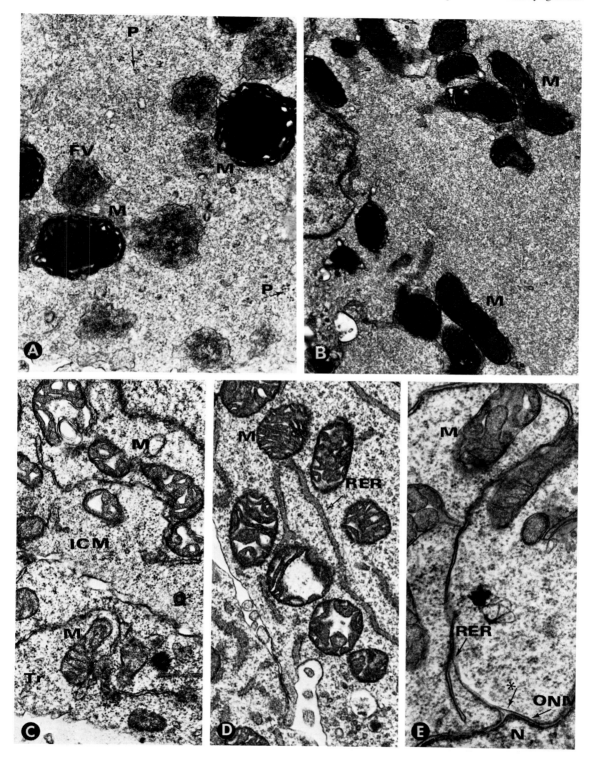

Plate 68. Mitochondrial Differentiation during Preimplantation Embryogenesis.

Plate 69. Membrane-Granule Complexes.

A   A rather characteristic feature of the cytoplasm of rabbit embryos during mid-cleavage are numerous "membrane-granule" complexes (MG) which frequently are observed in association with the nucleus (N). This micrograph also illustrates the appearance of a relatively mature nucleolus (nu). (x 6,000).

B   At higher magnification, the membrane-granule complexes (MG) are observed to be composed of whirls of membranes in which small, electron-dense granules are located. (x 15,800).

216

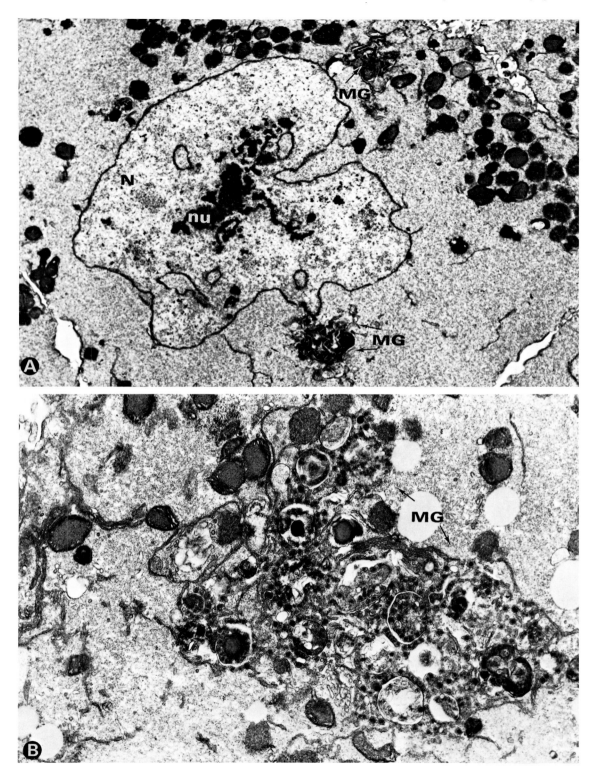

Plate 69. Membrane-Granule Complexes.

Plate 70. Membrane-Granule Complexes.

A   Transmission electron micrograph of a late rabbit morula demonstrating the periodic nature of the granules present within the complex and an apparent structural continuity between flocculent vesicles (FV) and the complex (arrows). It is not readily apparent from this micrograph whether flocculent vesicles are in fact continuous with the outer nuclear membrane. However, such continuity is often observed, suggesting to some investigators that the flocculent vesicles are derived (in part?) from the nucleus (see Plate 71C). N = nucleus; nu = nucleolus. (x 27,000).

B   Frequently, large, electron-dense lipid droplets (L) are encountered within the membrane-granule complexes (MG). Whether this material is broken down by the complex (resulting in the periodicity of the smaller, membranous-enclosed granules) or is in fact formed by the coalescence of the smaller dense bodies present within the complex is unknown. Also associated with the complexes are small, distended saccules of the smooth-surfaced endoplasmic reticulum (arrows). These saccules are observed to underlie the plasma membrane, but whether they originate from the complexes is not clear. The function(s) of the membrane-granule complexes observed in rabbit morulae is obscure, although they may be involved in the processing of material contained within the flocculent vesicles or in the production and storage of material to be required by the blastocyst for the rapid generation of plasma membranes. (x 15,000).

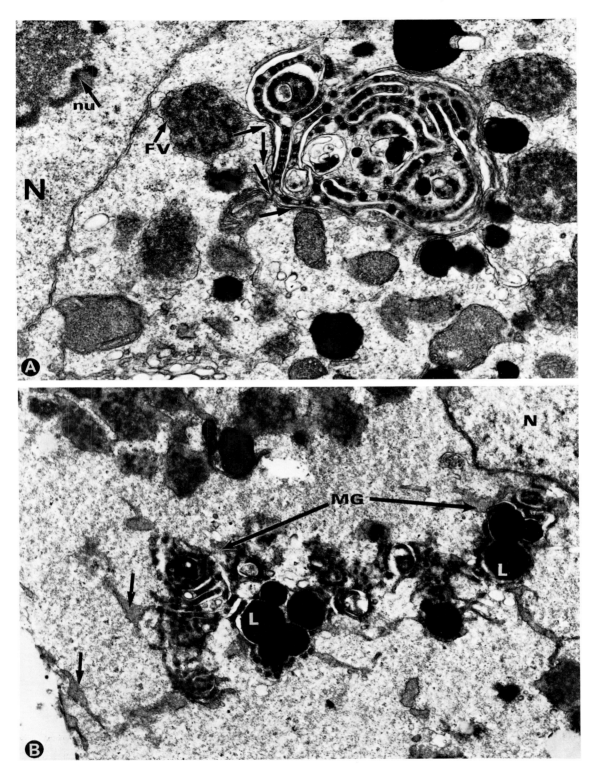

Plate 70. Membrane-Granule Complexes.

Plate 71. Flocculent Vesicles.

A   Freeze-fracture view of a trophoblast cell in an early cavitating rabbit embryo. Of particular interest is the appearance of the nucleus (N) and juxtanuclear Golgi complex (G). The numerous large indentations may represent mitochondria or residual flocculent vesicles. (x 16,000).

B   Flocculent vesicles in cleaving rabbit embryos frequently display a central, relatively clear area containing one or more small, electron-dense lipid-like droplets (arrows). The periphery of the vesicle is composed of an amorphous material possessing a granular texture. (x 14,600).

C   In serial-sectioned rabbit embryos, a structural continuity between the limiting membrane of the flocculent vesicle (FV) and the nuclear envelope (N) is often observed. In this micrograph, the space between the inner and outer nuclear membranes ("perinuclear cisternae") appears to be distended by an amorphous material that has the same texture as the material composing the flocculent vesicles (arrow). This apparent continuity has suggested to some investigators that the flocculent vesicles are derived from the nucleus. (x 36,400).

D   At the periphery of the cytoplasm of cleavage-stage blastomeres, flocculent vesicles, some containing a central electron-dense body, are typically associated with small saccules. These saccules are distended by a substance that has the same morphological texture as the material contained within the flocculent vesicles (arrow). It is unknown whether these small saccules are (1) fragmentation products of the flocculent vesicles, (2) extensions of the vesicles which in thin section appear as separate entities, or (3) actually contribute to the development of the vesicles. (x 30,000).

220

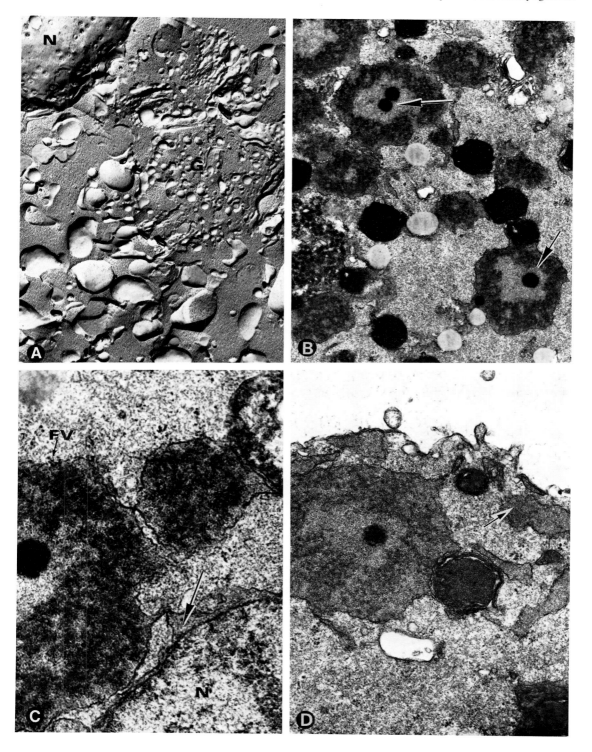

Plate 71. Flocculent Vesicles.

Plate 72. Intercellular Associations during Preimplantation Development.

A   During the early stages of cleavage in the rabbit, intercellular contact is maintained by means of cellular projections (arrows). N = nuclei; nu = nucleoli. (x 2,200).

B   At higher magnifications, the contacts between cellular projections of early-morula-stage rabbit embryos appear to be maintained primarily by means of gap-type junctions (J). The characteristic intercellular space between two adjacent blastomeres is indicated by the arrow. (x 25,000).

C   Scanning electron microscopic stereo-pair of a two-cell mouse embryo in which the zona pellucida has been removed enzymatically. Notice in particular that, with the exception of the area of immediate cell-to-cell contact, the entire surface of the cells is covered by a dense population of microvilli. (x 1,300).

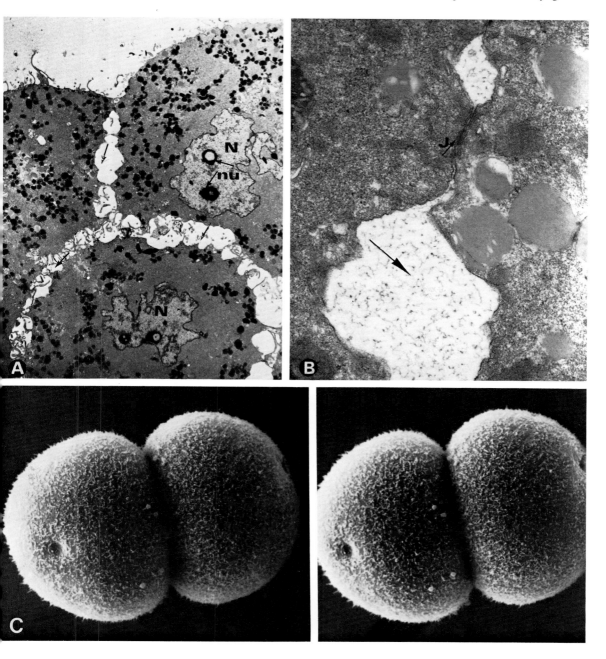

Plate 72. Intercellular Associations during Preimplantation Development.

Plate 73. Early Mouse Morulae.

A The appearance of a four-cell mouse embryo is shown in this transmission electron micrograph. Of particular interest are the fairly extensive intercellular spaces and the fine structure of the nucleoli (nu). Notice that the nucleolus in the lower blastomere is somewhat more advanced in its maturation than are the nucleoli in the adjacent blastomere. In other sections of this particular embryo, the dimensions of intercellular spaces were significantly smaller, and more closely resembled the appearance of the early rabbit morula shown in Plate 72A. (x 2,200).

B Scanning electron microscopic stereo-view of a precompaction mouse morula. Note that the outlines of each blastomere are quite distinct and that microvilli cover much of the cell surface. However, in regions where the blastomeres are in close apposition, the density of microvilli decreases markedly (*). (x 1,450).

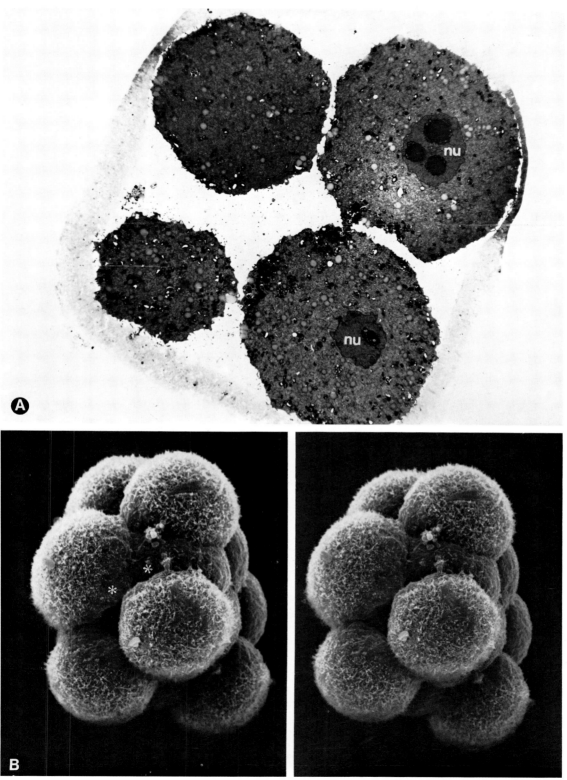

Plate 73. Early Mouse Morulae.

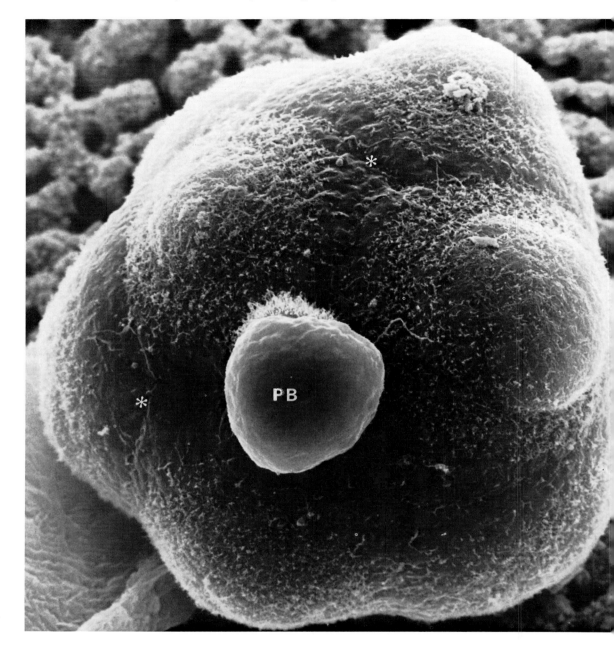

Plate 74. Compacted Mouse Morula.

    Scanning electron micrograph of a compacted mouse morula. Note the loss of distinct cellular outlines, the apparent restriction of microvilli to the more apical regions of the blastomeres, and the comparative lack of microvilli in regions of direct intercellular contact (*). Compaction is a morphogenetic process that involves the maximization of intercellular contact between the blastomeres. This contact is maintained to a large extent by the development of zonular tight junctions. PB = polar body. (x 4,000).

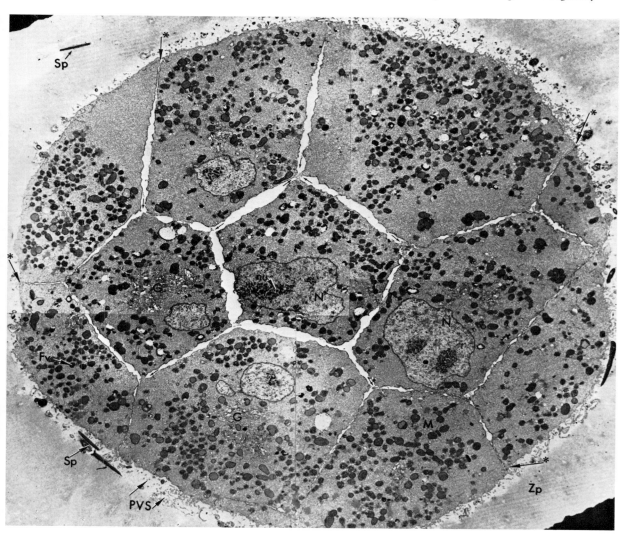

Plate 75. Rabbit Morula.

This montage of transmission electron micrographs demonstrates the characteristic appearance of a late-morula-stage rabbit embryo. Notice that spermatozoa (Sp) are still present both in the zona pellucida (Zp) and in the perivitelline space (PVS). Prominent nucleoli (nu) and juxtanuclear Golgi complexes (G) are evident. Although fairly large intercellular spaces separate the blastomeres, the apical portions of adjacent cells are interconnected by means of tight junctions (arrows*) which establish an effective permeability seal between the interior of the embryo and the milieu of the female reproductive tract. (x 2,200).

227

Plate 76. Surface Aspects of Precompaction Mouse Morulae.

A High-magnification scanning electron micrograph of a precompacted mouse morula. Note the sharp interface between a microvillous region (arrow*) and a zone containing few microvilli (arrow). The absence of a dense population of microvilli is characteristic of the portion of the cell membranes that are directly apposed. (x 7,000).

B Scanning electron micrograph demonstrating a differential distribution of microvilli between two apposing blastomeres in an early eight-cell mouse embryo. Notice the relative lack of microvilli on the surface of the cell on the left (arrow). This situation is typical of an early eight-cell embryo, but during the eight-cell stage, microvilli tend to disappear from all regions of intimate intercellular contact (see A). (x 7,000).

C At higher magnification, the borders between two apposing blastomeres in a precompaction mouse embryo display few microvilli. However, those that are present tend to be located in the intercellular zone (arrows). Frequently, elongated microvilli from one blastomere come to lie on the surface of the adjacent blastomere (*). This arrangement tends to bring the two cells into closer apposition. (x 13,000).

228

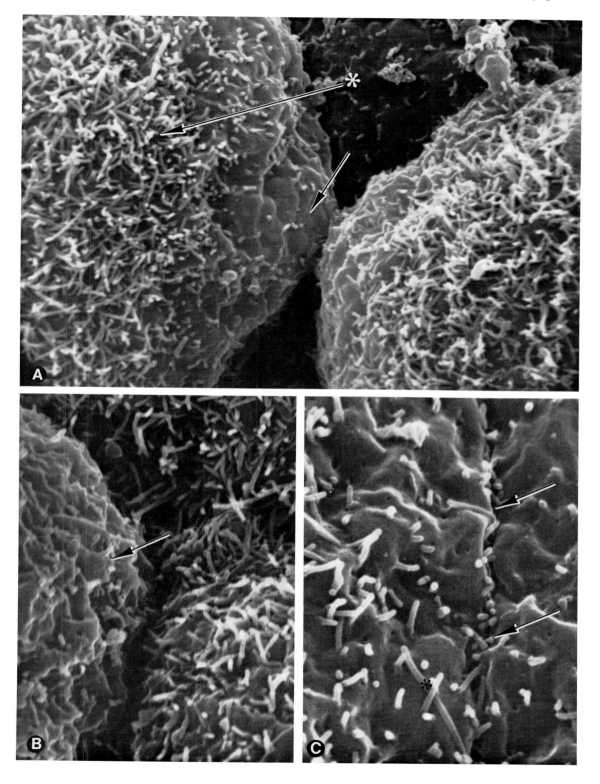

Plate 76. Surface Aspects of Precompaction Mouse Morulae.

Plate 77. Early Cavitating Mouse Blastocyst.

Transmission electron micrograph of an early cavitating mouse blastocyst ("signet-ring stage"). Of particular interest in this micrograph are the presence of a blastocyst cavity (BC) and the appearance of trophoblast cells (Tr). Trophoblast cells can become extremely thin (Tr-arrow), while by contrast, the cells of the inner cell mass (ICM) remain rounded. Also noteworthy are the presence of intercellular spaces between inner cell mass cells and the morphology of the nucleoli which are in an advanced stage of maturation (nu). N = nuclei; Zp = zona pellucida. (x 3,800).

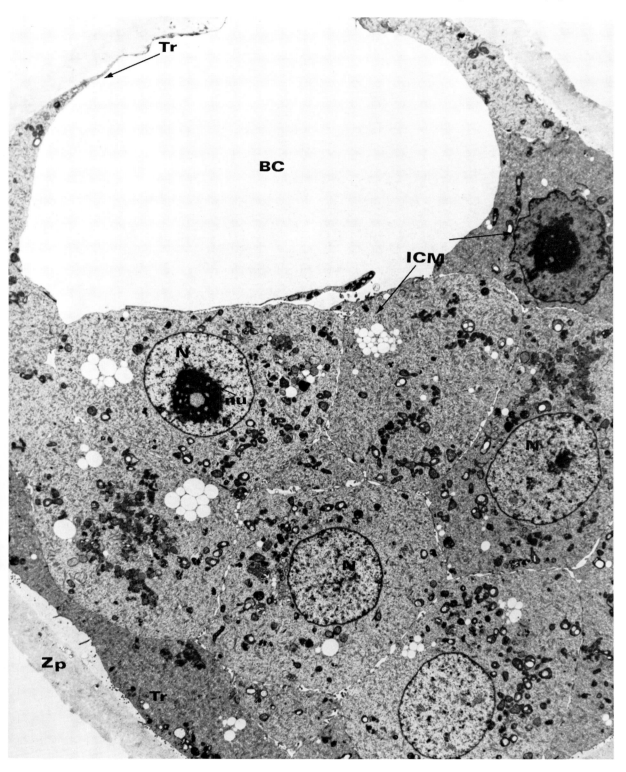

Plate 77. Early Cavitating Mouse Blastocyst.

Plate 78. Inner Cell Mass and Trophoblast of a Rabbit Blastocyst.

A   In early rabbit blastocyst, trophoblast cells usually have a columnar form. Typically, microvilli (Mv) densely populate the surface of the trophoblast that faces the zona pellucida (Zp). Notice the rather advanced stage of maturation attained by the nucleoli (nu). N = nuclei. (x 4,400).

B   During the stage of rapid blastocyst expansion, the trophoblast cells (Tr) in a rabbit embryo become extremely thin, while the cells of the inner cell mass (ICM) remain rounded. Again, notice the differential distribution of microvilli between the surface of the trophoblast that faces the blastocyst cavity (BC) and that which faces the zona pellucida (Zp). (x 2,500).

C   The mitotic division of trophoblast cells occurs in such a manner that while one cell faces the blastocyst cavity (BC), the sister cell faces the exterior of the embryo. However, intercellular contact is maintained during cell division by the establishment of zonular tight junctions. Notice the polarity of microvilli (Mv) between the region of the trophoblast facing the blastocyst cavity and that which faces the reproductive tract. One of the prominent features of mitosis in trophoblast cells is the presence of a midbody (arrow), usually at the lateral borders of recently divided cells. (x 3,500).

D   At higher magnification, the characteristic structure of the midbody is apparent. Midbodies contain clusters of microtubules that radiate into the cytoplasm and a band of high electron density. Notice the presence of zonular tight junctions between portions of the cytoplasm of the two sister cells (J). (x 25,000).

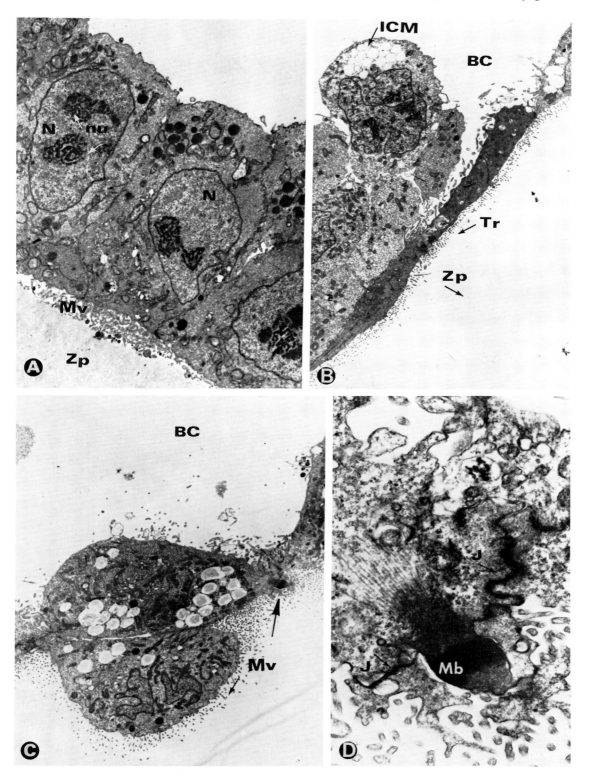

Plate 78. Inner Cell Mass and Trophoblast of a Rabbit Blastocyst.

Plate 79. Intercellular Junctional Complexes in Rabbit Trophoblast.

A   During the initial phases of blastocyst expansion in the rabbit, intercellular contact between trophoblast cells is maintained by tight junctions (Tj) located at the apical margins of the cells and by early or forming desmosomes present at the lateral borders (J). Early or forming desmosomes contain an electron-dense intercellular material and small aggregates of filaments (arrow) which run parallel to the plasma membrane on either side of the dense material. (x 43,000).

B   With continued embryonic expansion, characteristic desmosomes (D) appear at the lateral borders of adjacent cells, while zonular tight junctions (Tj) are present at the apical zones. BC = blastocyst cavity. (x 31,000).

C   Elaborate desmosomes (D) associated with filaments (F) that (1) originate from deep within the cytoplasm and converge at the plasma membrane, and (2) are oriented parallel to the plasma membrane, are typical of trophoblast cells during the expanded blastocyst stage. Desmosomes serve to provide the necessary internal structure required by the cells to withstand the hydrostatic pressures generated during cavitation. G = Golgi complex; C = centriole. (x 34,000).

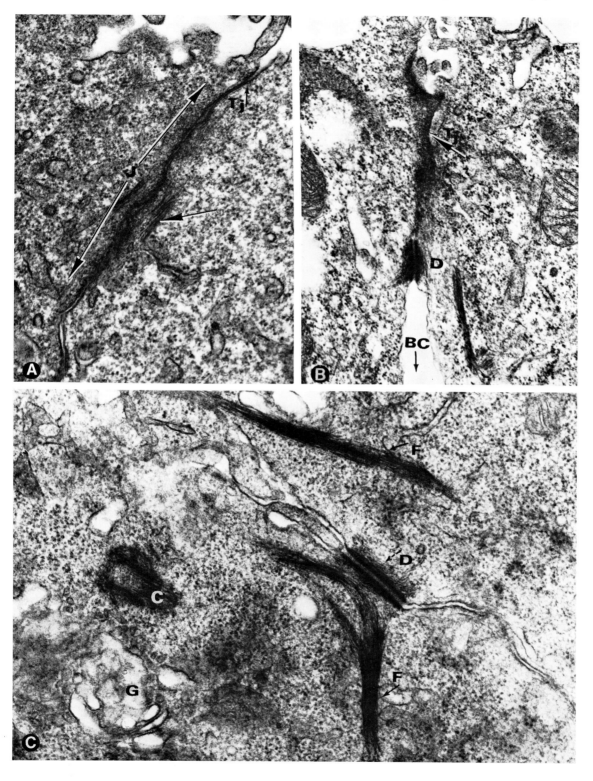

Plate 79. Intercellular Junctional Complexes in Rabbit Trophoblast.

Plate 80. Intercellular Junctions.

A   The circumferential distribution of desmosome-associated cytoplasmic filaments (F) is partially evident in this transmission electron micrograph of an expanded rabbit blastocyst. N = nucleus; L = accumulation of lipid droplets; Mv = microvilli; BC = blastocyst cavity. (x 5,700).

B   Transmission electron micrograph of extremely laminar trophoblast cells in a 6-day-old rabbit embryo. Note the series of desmosomes between two trophoblast cells (D) and the polarity in the distribution of microvilli (Mv) between the interior and exterior surfaces of these cells. BC = blastocyst cavity. (x 27,000).

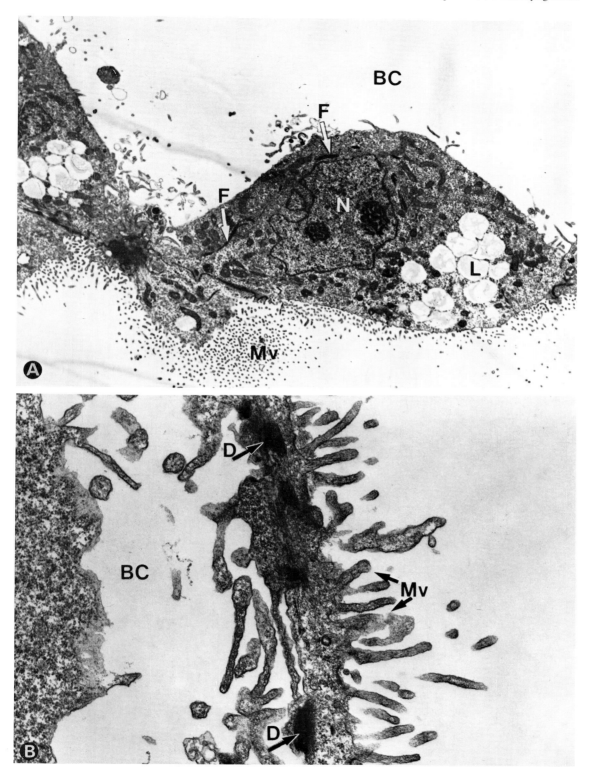

Plate 80. Intercellular Junctions.

Plate 81. Surface View of an Expanded Rabbit Blastocyst.

A   In this scanning electron micrograph of an expanded rabbit blastocyst, the rather thin nature of both the zona pellucida (Zp) and the trophoblast layer (Tr) is apparent. Accumulations of an amorphous material are a characteristic feature of the surface of the zona pellucida (*). (x 2,500).

B   In this view of an expanded rabbit blastocyst, the microvillous (Mv) appearance of the surface of the trophoblast (Tr) that faces the zona pellucida (Zp) is evident. In addition, the zona has a textured surface. (x 2,500).

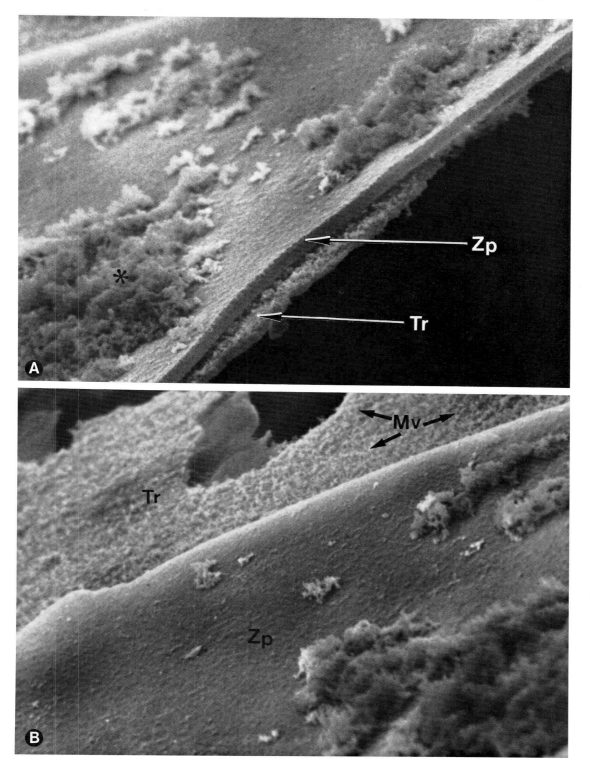

Plate 81. Surface View of an Expanded Rabbit Blastocyst.

Plate 82. Surfce Views of Blastocyst-Stage Embryos.

A   The interior surface of the trophoblast of a rabbit blastocyst is shown in this scanning electron micrograph. While the surface of the trophoblast that faces the zona pellucida is densely covered by microvilli (Plates 81B, 83B), the interior membrane of the trophoblast adjacent to the blastocyst cavity displays a reduced population of microvilli and a central area or bulge that corresponds to the region of the cytoplasm occupied by the nucleus (N). (x 1,800).

B   Scanning electron microscopic stereo-pair of a hatched mouse blastocyst. Notice that the trophoblast cells are rounded and that the surfaces of these cells display a series of ridges (arrows). (x 1,300).

240

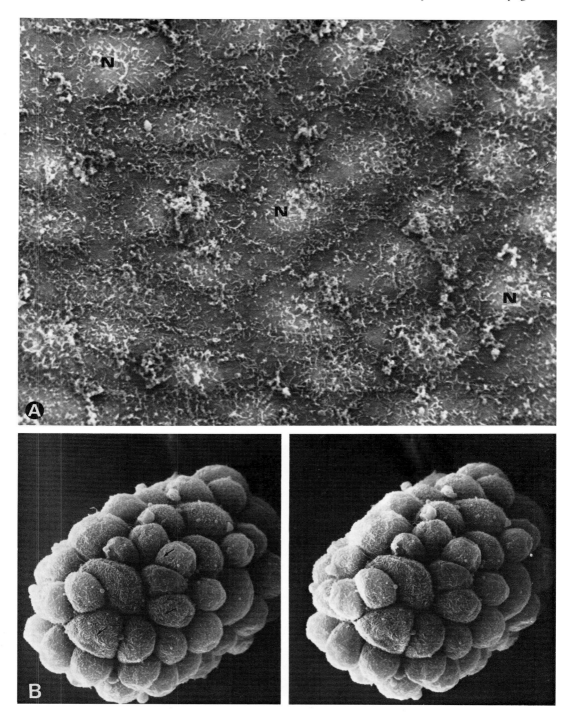

Plate 82. Surface Views of Blastocyst-Stage Embryos.

Plate 83. Surface Views of Blastocyst-Stage Embryos.

A   A typical surface feature of late mouse blastocysts is the presence at regions of intercellular contact of Y-shaped structures (arrows). It is not known whether these structures are involved in intercellular communication or are a surface manifestation of the underlying junctional complexes. (x 18,000).

B   Surface view of a trophoblast cell in a fully expanded rabbit blastocyst (day 6.5 following fertilization). Trophoblast cells are quite flattened, and display numerous microvilli (Mv) and a central bulge corresponding to the region of the cytoplasm occupied by the nucleus (N). Zp = zona pellucida. (x 11,000).

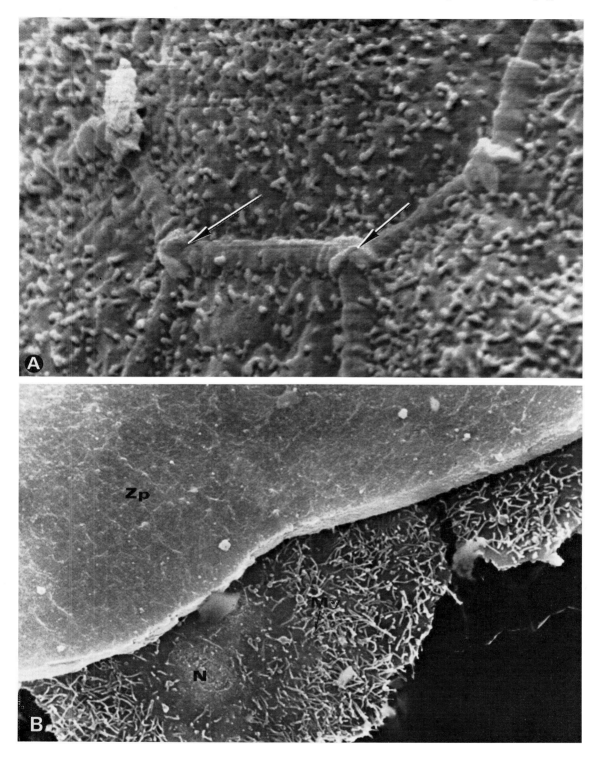

Plate 83. Surface Views of Blastocyst-Stage Embryos.

Plate 84. Inner Cell Mass and Endoderm in the Rabbit Blastocyst.

A   As observed by scanning electron microscopy, the internal surface of the inner cell mass is composed of cells that have distinct cellular borders. Few microvilli are expressed on these cells. (x 1,100).

B   Scanning electron microscopic view of parietal endodermal cells (En) in the process of migrating over the interior surface of the trophoblast (Tr). Although large intercellular spaces separate endodermal cells, these cells maintain contact with one another by means of extended cytoplasmic projections (arrows). (x 6,000).

C   Where endodermal cells are in intimate association, the borders of adjacent cells are observed to contain numerous interdigitations (arrow) of cytoplasmic processes, as well as small intercellular spaces (*). (x 16,000).

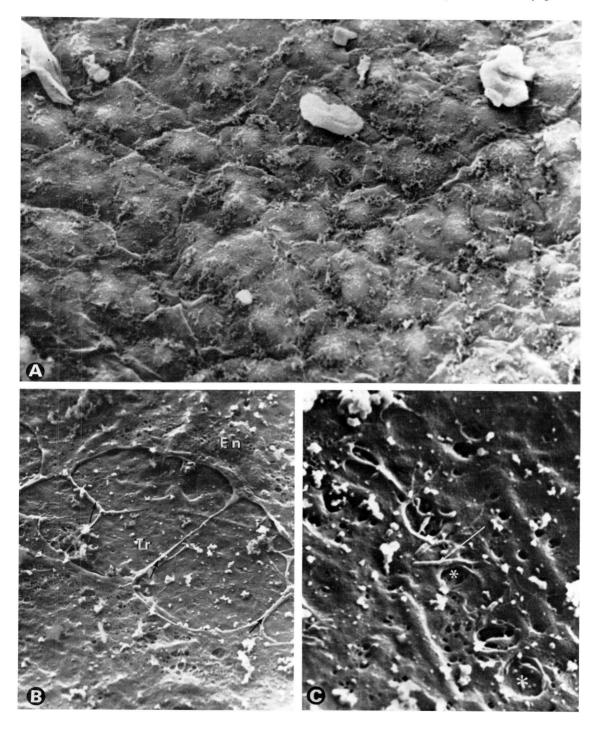

Plate 84. Inner Cell Mass and Endoderm in the Rabbit Blastocyst.

Plate 85. Endodermal Cells in Fully Expanded Preimplantation Rabbit Blastocysts.

A Occasionally, endodermal cells, probably in the process of mitosis are observed by scanning electron microscopy. These particular cells are recognized by (1) their rounded appearance, (2) the presence of numerous surface blebs (large arrows), and (3) the elaboration of a network of peripheral cytoplasmic projections (filapodia) which anchor onto the surface of the overlying trophoblast (smaller arrows). (x 8,500).

B Contact between the cellular projections of two endodermal cells is demonstrated in this scanning electron micrograph. In addition, the surface of the trophoblast (Tr) which lines the blastocyst cavity contains small accumulations of an amorphous material. This material is also observed in the external surface of endodermal cells (see A). (x 14,000).

246

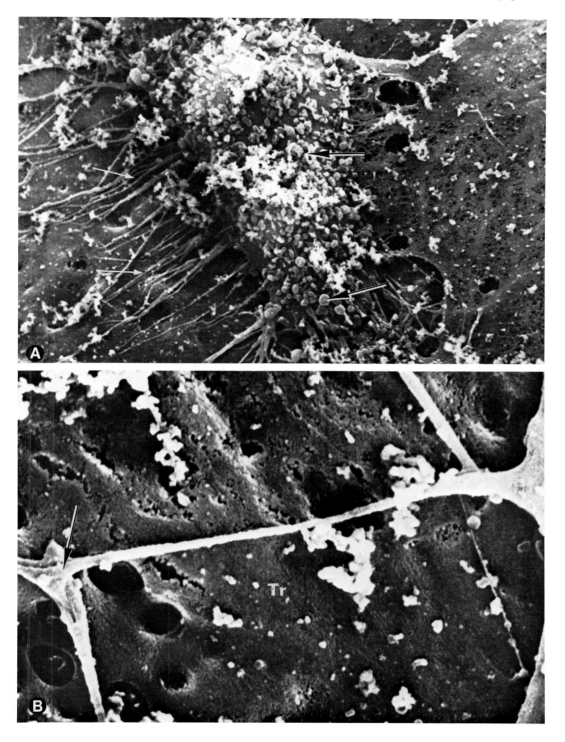

Plate 85. Endodermal Cells in Fully Expanded Preimplantation Rabbit Blastocysts.

Index

Note: Numerals in roman type refer to pages; numerals in italic type refer to plate numbers.